Grade 5
Mathematics
Student Book 5B

PAT LILBURN
PETER SULLIVAN

in consultation with
Elsie Kinavai
Alex Feeger

Department of Education Papua New Guinea

OXFORD
UNIVERSITY PRESS
AUSTRALIA & NEW ZEALAND

OXFORD
UNIVERSITY PRESS
AUSTRALIA & NEW ZEALAND

253 Normanby Road, South Melbourne, Victoria 3205, Australia

Oxford University Press is a department of the University of Oxford.
It furthers the University's objective of excellence in research,
scholarship, and education by publishing worldwide in

Oxford New York

Auckland Cape Town Dar es Salaam Hong Kong Karachi
Kuala Lumpur Madrid Melbourne Mexico City Nairobi
New Delhi Shanghai Taipei Toronto

With offices in

Argentina Austria Brazil Chile Czech Republic France Greece Guatemala
Hungary Italy Japan Poland Portugal Singapore South Korea Switzerland
Thailand Turkey Ukraine Vietnam

First published 2001
Reprinted 2003, 2008 (twice), 2009, 2010

ISBN 978 0 19 550866 6

Cover design by Sylvia Witte
Typesetting by Promptset Pty Ltd (Stephen Chan)
Cover artwork by Gigs Wena
Illustrated by Ophelia Leviny, Boris Sylvestri and Promptset Pty Ltd
Printed in China by Golden Cup Printing Co. Ltd
Published by Oxford University Press
Editorial Office: PO Box 7979, Boroko NCD, Papua New Guinea

Secretary's Message

This student book is part of the new reformed mathematics program designed and written for use in Grade 5 classes in Community ar Primary schools throughout Papua New Guinea.

The core materials consist of two student books called *Grade 5 Mathematics Student Book 5A* and *Grade 5 Mathematics Student Book 5B*. They are accompanied by the *Grade 5 Mathematics Teacher's Resource Book*. They replace the MACS series now in use.

In the Mathematics program students are first taught mathematics by using real objects. Later, the students will use pictures of objects. Finally, the students will use number symbols to represent these objects. Always allow your students to use real objects durin their mathematics lessons if they want to. Teach the concepts in the context of real life situations as this leads to an appreciation of th everyday use of mathematics skills and knowledge. The students will decide when they do not need the help of real objects any longe Remember, when students use real objects, they will understand mathematics better. This program also encourages the students to solve their own problems and make their own decisions with confidence. In order to learn these skills, the students should talk about what they are doing in every lesson. Since learning is most effective when it has meaning and is enjoyable, allow the students to use the language that they are most comfortable with.

However, in Grade 5 more English should be used together with the vernacular and other languages to aid understanding of new and difficult concepts.

Finally, the National Department of Education wants teachers to be flexible in programming and timetabling to cater for the different abilities of students and multigrade teaching situations. This book shows you some strategies to help satisfy these needs.

Peter M. Baki

PETER M. BAKI
Secretary of Education

CONTENTS

Which way will it roll?

Place a counter in the top circle. Toss a coin. If it lands on heads, move the counter left. If it lands on tails, move it right. Toss the coin at each circle until the counter reaches the boxes at the bottom. Do this 30 times. Where are the counters most likely to finish?

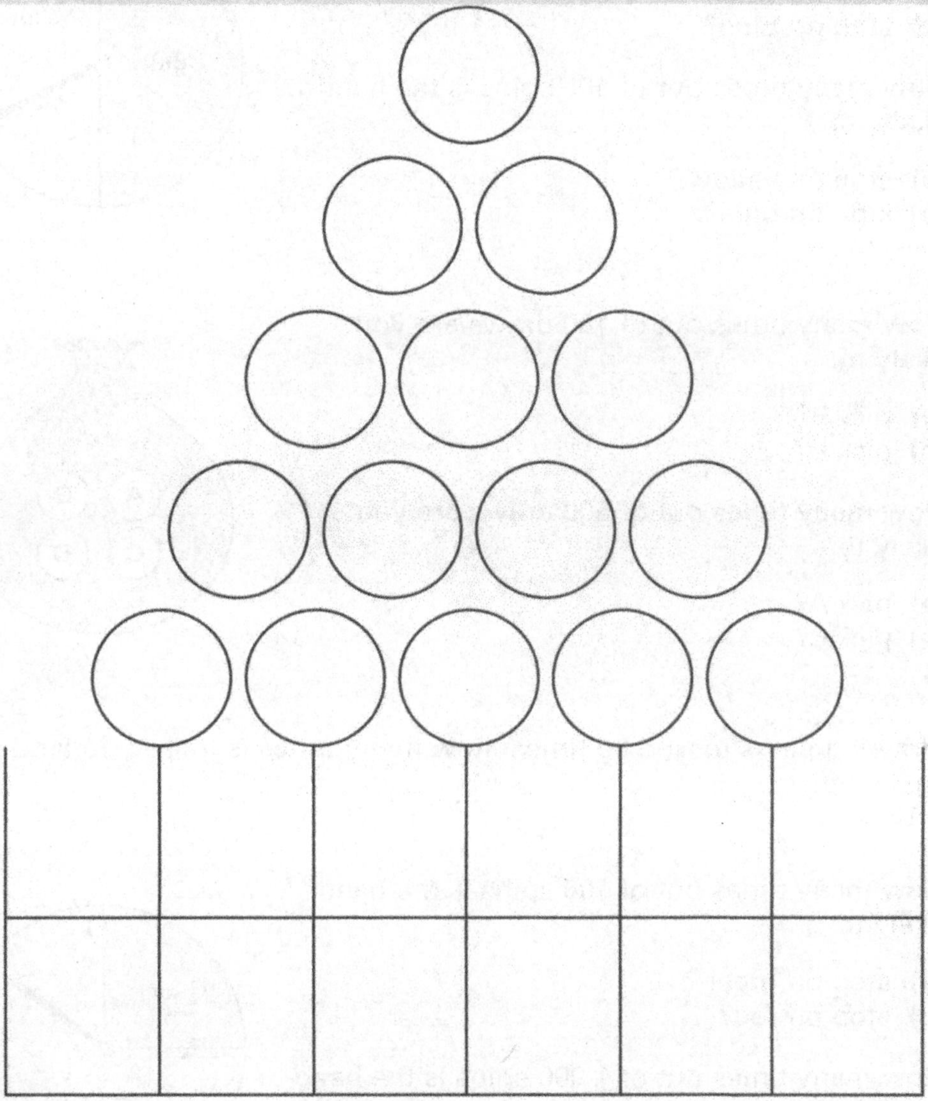

What are the chances?

A

1. How many times out of 100 spins is the hand likely to:

   (a) stop on yellow?
   (b) stop on blue?

2. How many times out of 500 spins is the hand likely to:

   (a) stop on yellow?
   (b) stop on blue?

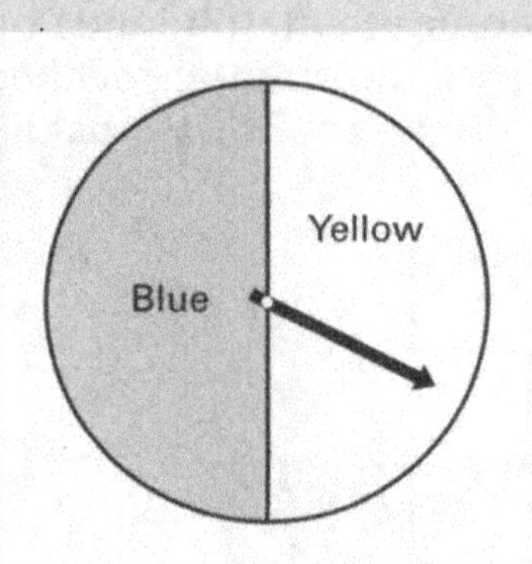

B

1. How many times out of 100 draws are you likely to:

   (a) pick B?
   (b) pick D?

2. How many times out of 500 draws are you likely to:

   (a) pick A?
   (b) pick C?

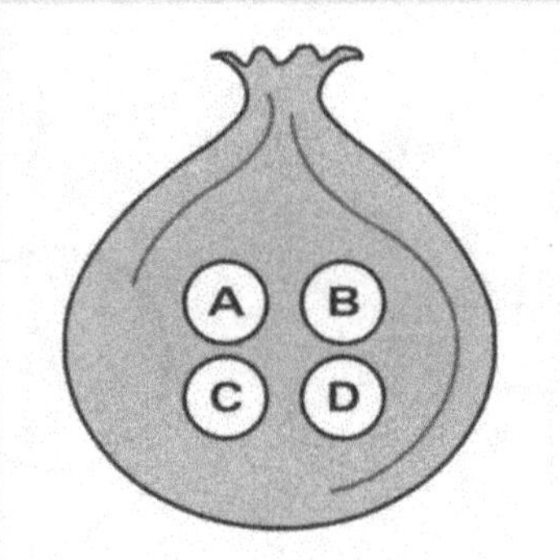

C

If a drawing pin is tossed 60 times, how many times is it likely to land with the spike up?

D

1. How many times out of 100 spins is the hand likely to:

   (a) stop on green?
   (b) stop on red?

2. How many times out of 1 000 spins is the hand likely to:

   (a) stop on green?
   (b) stop on red?

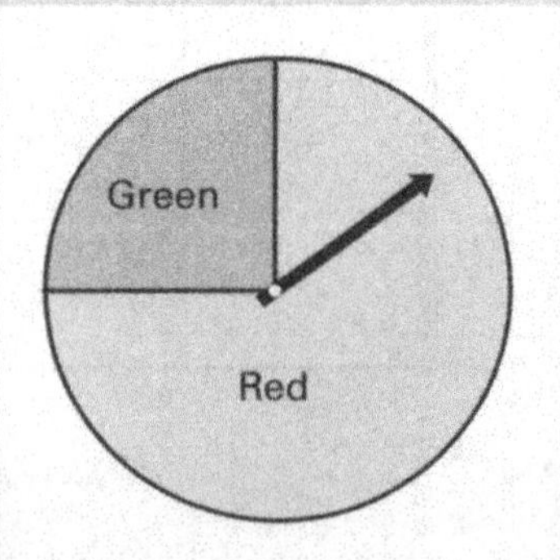

Exploring volume

Draw these nets onto centimetre squared grid paper, then cut them out and bend on the dotted lines. Tape them together.

Fill each shape with centimetre cubes to find the volume, then record it in your exercise book.

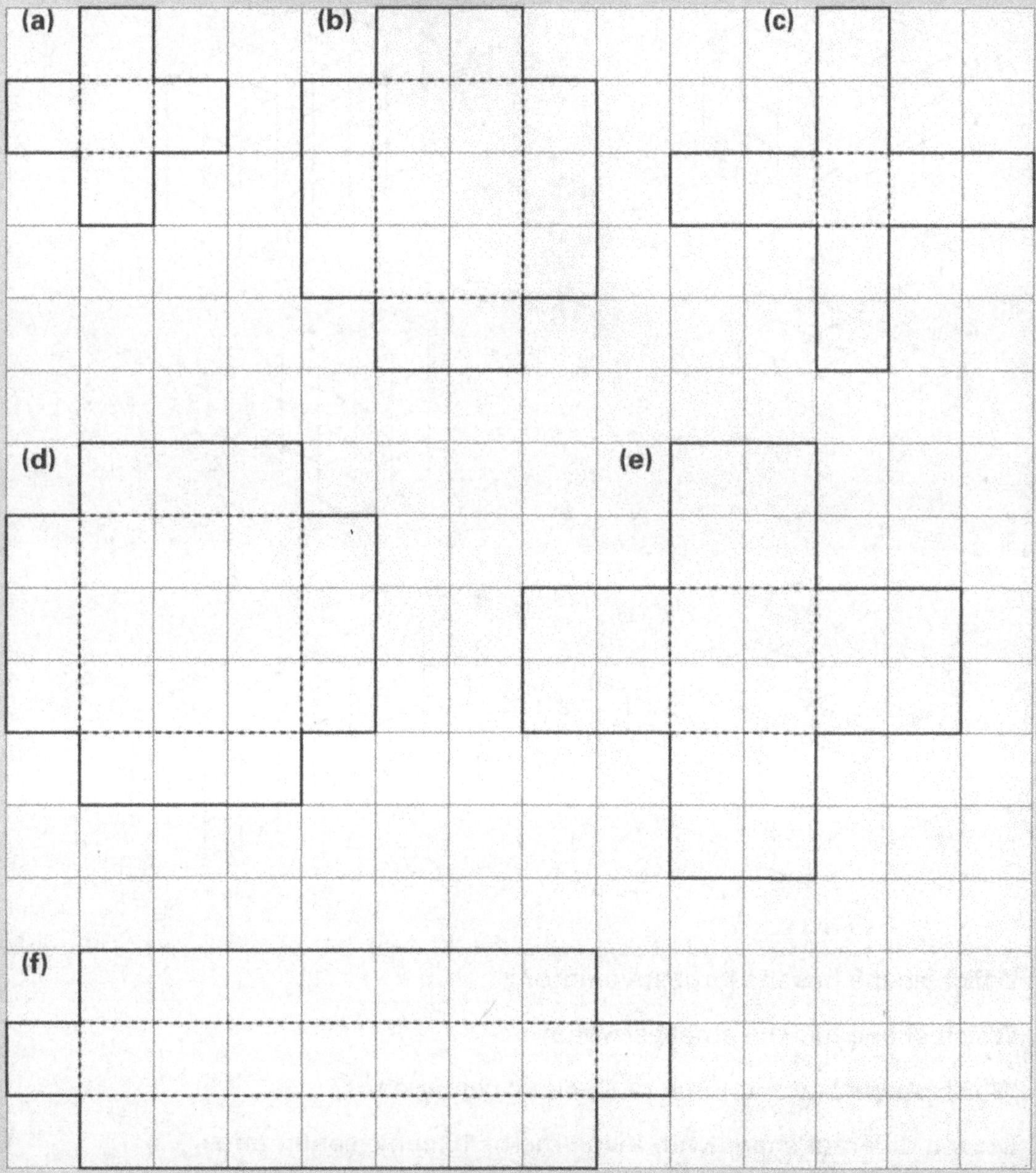

How many cubic centimetres?

What is the volume of each shape?

Write a multiplication number sentence to help you find the volume of each shape. Record your answers in cubic centimetres.

1.

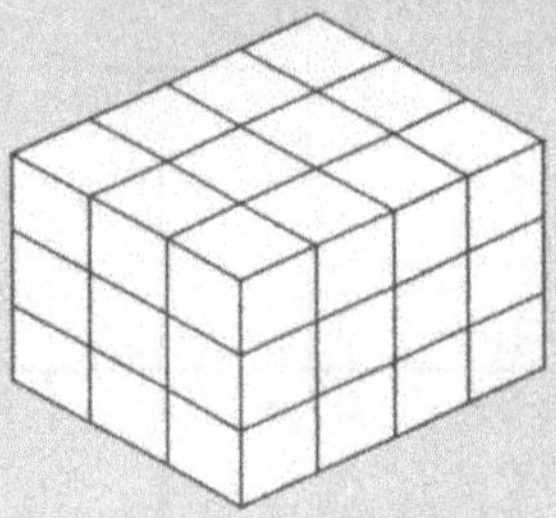

2.

3.

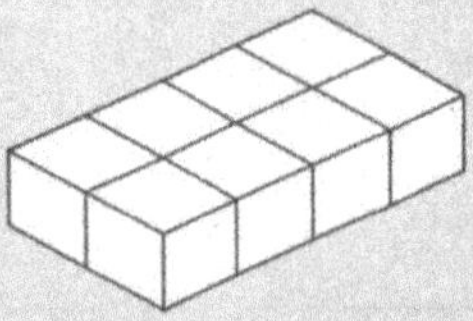

4.

5.

6.

Which shape has the largest volume?

Which shape has the smallest volume?

Which shape has a volume of 16 cubic centimetres?

Draw a different shape with a volume of 16 cubic centimetres.

What would you use?

1. What unit of volume would you use to measure these items?

(a)

(b)

(c)

(d)

(e)

(f)

2. Choose cubic centimetres or cubic metres to fill these spaces.

(a) garage: 75 ______ **(b)** loaf of bread: 7 000 ______

(c) packing crate: 2 ______ **(d)** brick: 1 600 ______

(e) pencil case: 960 ______ **(f)** matchbox: 20 ______

(g) classroom: 200 ______ **(h)** cupboard: 1 ______

(i) store room: 24 ______ **(j)** tool box: 10 000 ______

Finding volume of irregular solids

Each box is one cubic metre. Boxes have been stacked as shown. What is the volume of each stack?

A

B

C

D

E

F

G

1. What is the difference in volume between:

   (a) A and E? **(b)** F and G? **(c)** C and E? **(d)** B and G?
   (e) B and F? **(f)** D and C? **(g)** F and E? **(h)** A and B?

2. All the stacks are put together. How many cubic metres will there be?

How many cubic centimetres are not painted?

1. What is the volume of each solid?
2. Each solid is painted on the outside and then pulled apart. How many centimetre cubes have no paint on any surface?

A

B

C

D

E

Solving word problems

1. If a truck will hold about half a cubic metre of soil, how many trucks will be needed to transport 5 cubic metres?

2. Work out the differences in volume between the following solid shapes and write your answers in cubic centimetres.

   (a) 3 cm x 5 cm x 2 cm and 4 cm x 4 cm x 2 cm

   (b) 6 cm x 2 cm x 3 cm and 3 cm x 4 cm x 5 cm

   (c) 8 cm x 2 cm x 2 cm and 2 cm x 3 cm x 3 cm

3. A 200 litre tank is $\frac{3}{4}$ full of water. What is the volume of water in the tank?

4. When a water tank is $\frac{1}{4}$ full, it holds 15 L of water. How much water does it hold when it is full?

5. 12 crates, each with a volume of $\frac{1}{4}$ cubic metre, were loaded onto a truck. What was the total volume of the crates?

6. The maximum cargo allowance is 20 cubic metres. Which of A, B, C, D or E can I take?

   A $7\frac{1}{2}$ cubic metres and 3.25 cubic metres and 9 cubic metres.

   B 10 cubic metres and 4.5 cubic metres and 6.75 cubic metres.

   C $5\frac{1}{4}$ cubic metres and $12\frac{1}{4}$ cubic metres and 2 cubic metres.

   D 2.25 cubic metres and 10.5 cubic metres and 9 cubic metres.

   E $8\frac{3}{4}$ cubic metres and 5.25 cubic metres and 4.75 cubic metres.

7. We need 6 cubic metres of sand. Our truck will hold 1.5 cubic metres. How many trips will we need to make?

8. What is the volume of a concrete slab that is 60 cm wide, 80 cm long, and 6 cm high?

Capacity of fish tanks

Use the measurements given to work out the volume of water in each fish tank. You will need to convert cubic centimetres to millilitres.

Write your answers in litres and millilitres.

Remember: 1 cubic cm = 1 mL

1.

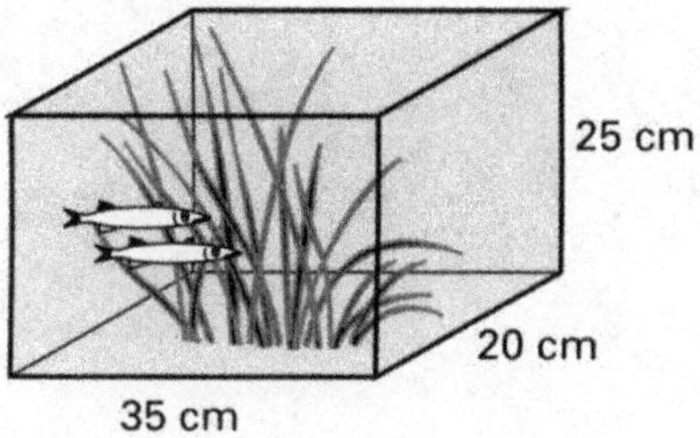

2.

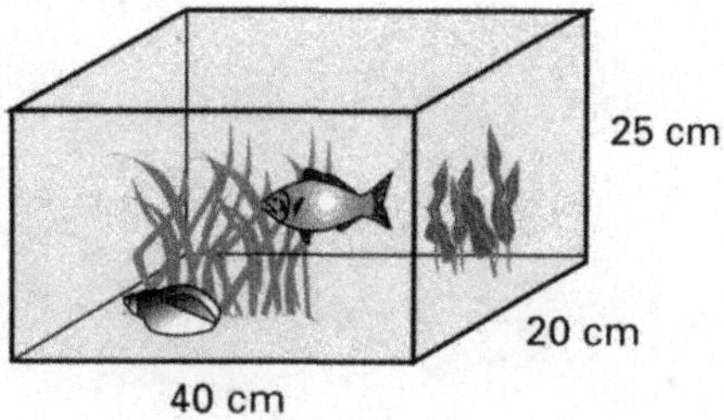

3.

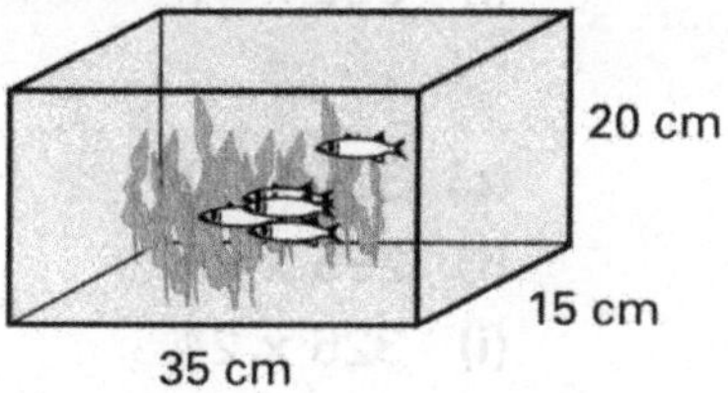

4.

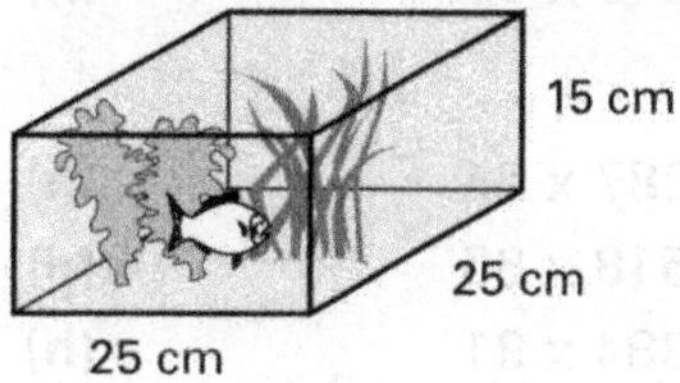

5.

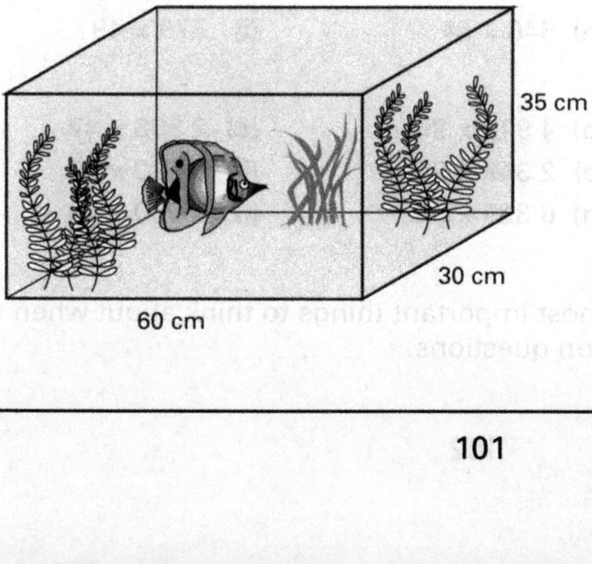

6.

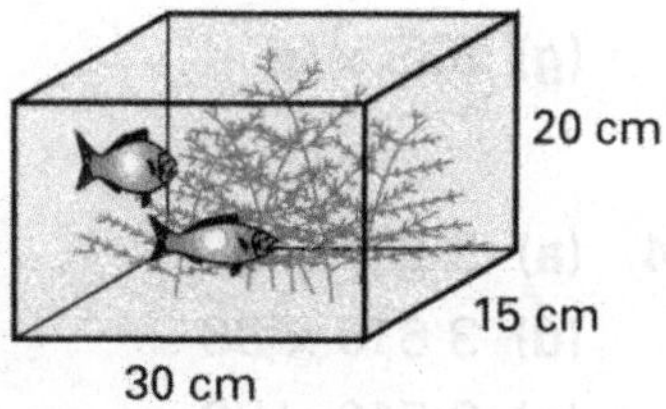

Long multiplication

Answer these questions in your book. You could record them as follows:

```
     1 2 1 1 2 5  6
×              3  4
   ----------------
           6   2  4
      1 4  1 6 8  0
   ----------------
        5  3   0  4
   ----------------
```

1. **(a)** 123 x 32 **(b)** 223 x 41 **(c)** 315 x 23
   (d) 812 x 24 **(e)** 513 x 32 **(f)** 421 x 14
   (g) 243 x 15 **(h)** 344 x 33 **(i)** 562 x 17

2. **(a)** 387 x 24 **(b)** 168 x 56 **(c)** 329 x 72
   (d) 518 x 86 **(e)** 792 x 77 **(f)** 295 x 83
   (g) 384 x 91 **(h)** 918 x 67 **(i)** 326 x 24

3. **(a)** 355 x 26 **(b)** 284 x 35 **(c)** 385 x 82
   (d) 468 x 45 **(e)** 285 x 62 **(f)** 594 x 53
   (g) 835 x 28 **(h)** 846 x 54 **(i)** 779 x 49

4. **(a)** 3 617 x 23 **(b)** 4 913 x 35 **(c)** 2 806 x 47
   (d) 3 619 x 38 **(e)** 2 349 x 24 **(f)** 3 600 x 36
   (g) 3 510 x 22 **(h)** 6 829 x 36 **(i)** 4 030 x 50

5. Write down what are the most important things to think about when you are doing long multiplication questions.

Multiplication story problems

Answer these questions in your book. Each of the questions has more than one possible correct answer. Give as many correct answers as you can for each question.

1. Two numbers are multiplied together to give 600. What could the numbers be?

2. Two numbers are multiplied together to give 2 280. What could the numbers be?

3. Using the digits 1, 2, 3, set out like ☐☐ x ☐ how many different answers can you get?

4. Using the digits 1, 2, 3 and 4 set out like ☐☐ x ☐☐ how many different answers can you get?

5. Using the digits 1, 2, 3 and 0 set out like ☐☐ x ☐☐ how many different answers can you get?

6. Using the digits 1, 2, 3 and 4 set out like ☐☐☐ x ☐ how many different answers can you get?

7. There are 240 students in a school. They are arranged into teams. The teams have the same number of people. How many teams could there be and how many people might there be in each team?

8. In this calculation some of the numbers are missing. What could the missing numbers be? 3☐ x 2☐ = ☐☐0.

9. In this calculation some of the numbers are missing. What could the missing numbers be? ☐3 x ☐2 = ☐☐6.

10. In this calculation some of the numbers are missing. What could the missing numbers be? ☐☐ x ☐☐ = ☐00.

Multiplication using decimals

Answer these questions in your book. You can set them out as follows:

$$\begin{array}{r} {}_{4}1.5 \\ \times \quad 9 \\ \hline 13.5 \\ \hline \end{array}$$

1. (a) 1.4 x 2 (b) 3.2 x 3 (c) 4.2 x 4
   (d) 5.3 x 3 (e) 7.2 x 4 (f) 8.6 x 6
   (g) 5.3 x 8 (h) 2.8 x 4 (i) 5.7 x 6

2. (a) 21.4 x 3 (b) 37.4 x 4 (c) 20.6 x 3
   (d) 82.7 x 6 (e) 38.3 x 5 (f) 60.4 x 9
   (g) 70.3 x 8 (h) 41.8 x 7 (i) 11.1 x 8

3. (a) 8.4 x 5 (b) 2.6 x 5 (c) 3.5 x 8
   (d) 4.5 x 6 (e) 4.5 x 4 (f) 8.5 x 2
   (g) 8.8 x 5 (h) 6.6 x 5 (i) 7.4 x 3

4. (a) 3.4 x 30 (b) 5.2 x 20 (c) 2.4 x 40
   (d) 6.3 x 50 (e) 2.4 x 60 (f) 3.6 x 30
   (g) 5.2 x 40 (h) 3.2 x 50 (i) 5.3 x 20

5. (a) 3.4 x 32 (b) 5.2 x 24 (c) 2.4 x 43
   (d) 6.3 x 52 (e) 2.4 x 62 (f) 3.6 x 34
   (g) 5.2 x 43 (h) 3.2 x 56 (i) 4.7 x 67

6. Write down what are the most important things to think about when you are multiplying decimals.

Multiplication story problems

Answer these questions in your book. Make sure that **you write a sentence to show the answer** to the question.

1. A church has seats in rows. There are 52 rows, and 157 seats in each row. How many people can sit in the church?
2. Three letters are used for car number plates. All letters can be used. How many different sets of three letters can be made?
3. We are planning a long trip by boat. We know that we will be able to catch fish along the way. We need to take some other food. We estimate that each week, each person will need:

   5 kg kaukau
   14 L water
   3 packets of biscuits
   2 kg rice
   5 kg fruit

   (a) How much kaukau will we need for 6 people for 4 weeks?
   (b) How much water would we need for 12 people for 6 weeks?
   (c) How many packets of biscuits would we need for 7 people for 4 weeks?
   (d) How much fruit would be needed for 14 people for 5 weeks?

4. We are planning to stock the school canteen. We estimate that each year, each student uses 3 exercise books, 4 pencils, 1 ruler, and 5 pens.

   (a) For a school of 240 pupils, how many exercise books will probably be used?
   (b) For a school of 371 pupils, how many pencils will probably be used?

Multiplication story problems using decimals

Answer these questions in your books. Make sure that you write a sentence to show the answer to the question.

1. A very small unit of weight called the milligram is used to measure medicine. A doctor prescribed medicine at a rate of 2.8 mg for each kilogram of body weight.

   (a) How much was needed for someone who weighed 57 kg?
   (b) How much was needed for someone who weighed 43 kg?
   (c) How much was needed for someone who weighed 82 kg?

2. Petrol is for sale at 180.5t per litre. How much would I pay for 52 litres?

3. I have 5 sheets of roofing iron. Each sheet is 0.7 m wide. What is the widest roof that I can cover with these sheets?

4. Chicken wire is 0.9 m high. If I put three layers one above the other, what is the height of the fence?

5. Lucy wants to make herself a dress. She will need 3 m of material. The material she likes costs K3.70 a metre. How much will it cost her to buy the material?

Time calculations

Answer these questions in your book.

1. How many days are there in:

   (a) 5 years
   (b) 27 years
   (c) 74 years
   (d) 3 decades
   (e) 2 centuries
   (f) 100 years?

2. How many days are there in:

   (a) 48 hours
   (b) a week
   (c) this month
   (d) next month
   (e) this year
   (f) a fortnight?

3. How many minutes are there in:

   (a) an hour
   (b) a day
   (c) this week
   (d) this month
   (e) the next two months after this month
   (f) last week?

4. How many seconds are there in:

   (a) an hour
   (b) a day
   (c) a week
   (d) this month
   (e) last month
   (f) next month?

5. How far would a plane travelling at 350 km/hr travel in 6 hours?

6. How far would a plane travelling at 720 km/hr travel in 4 hours?

7. How far would a plane travelling at 285 km/hr travel in 15 hours?

8. How far would a plane travelling at 770 km/hr travel in 13 hours?

Counting all the pieces

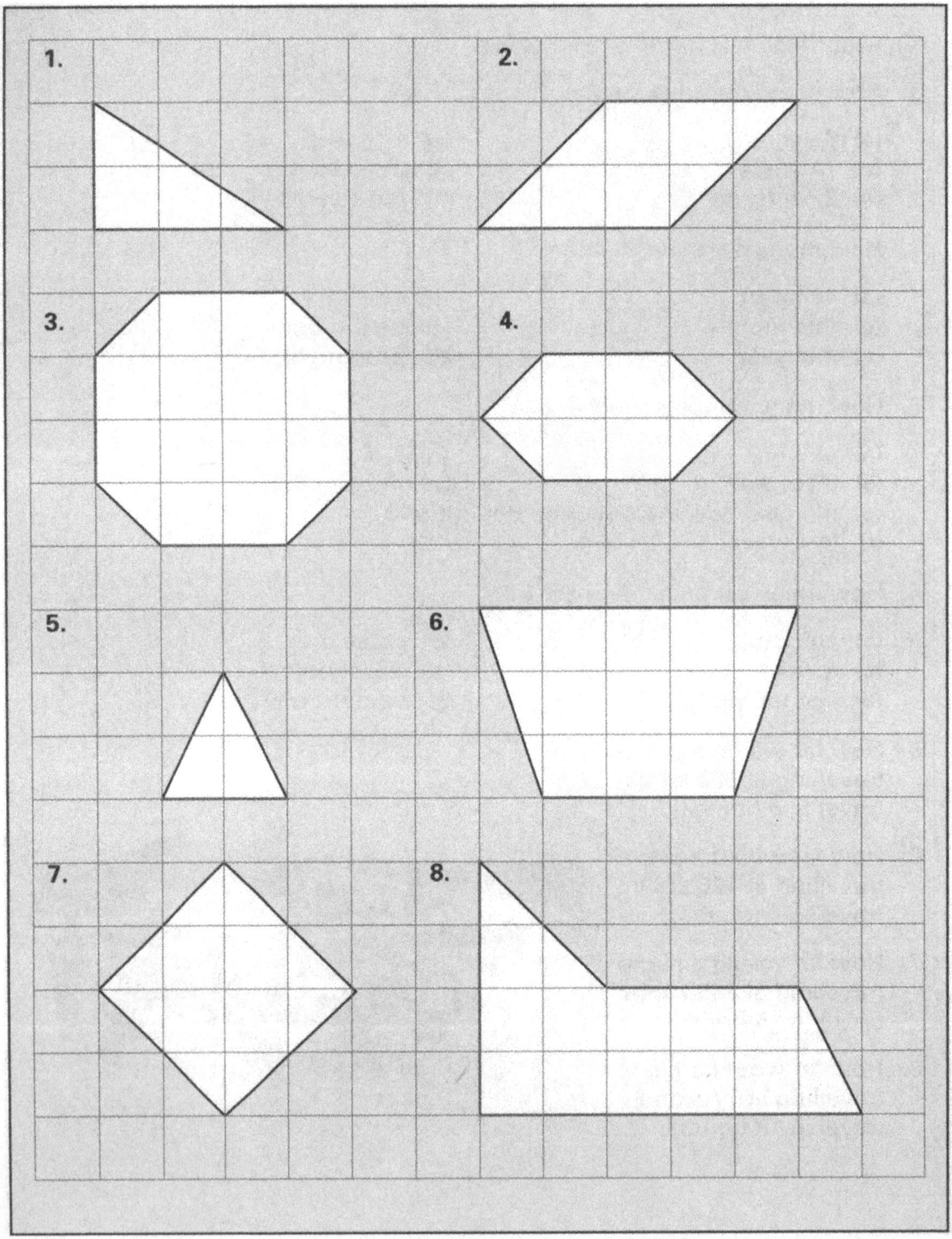

A short cut for calculating area

1. Each of the small squares in this diagram represents a square that is 1 m long and 1 m wide. What is the area of each of the shaded rectangles?

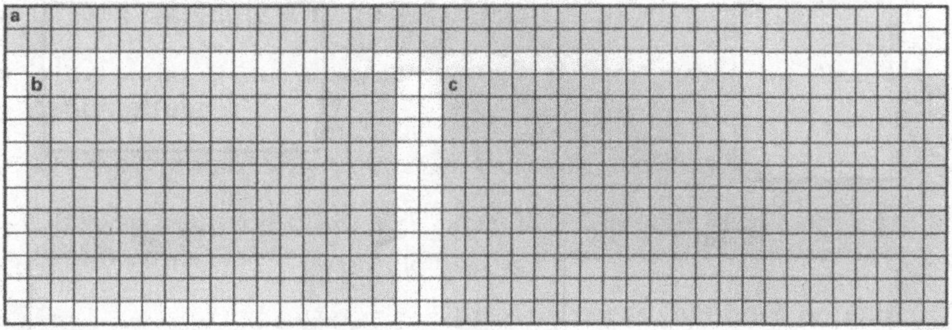

2. How can you work out the area of the shapes in question 1 without counting every square?

3. How many small squares are needed to fill each of the following rectangles?

(a)

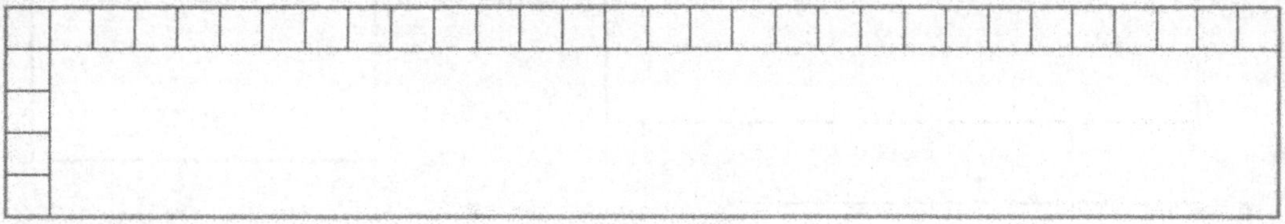

(b)

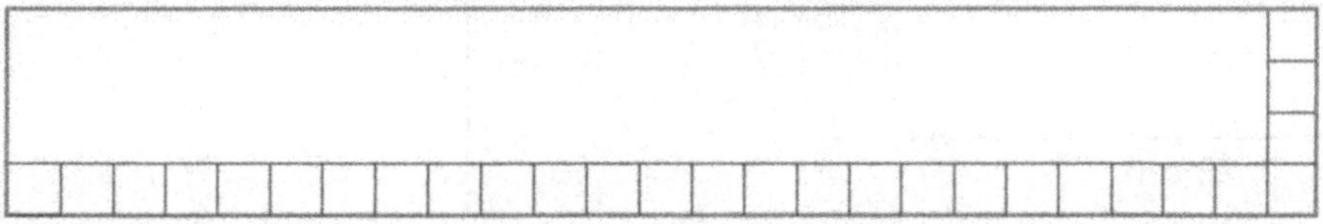

(c)

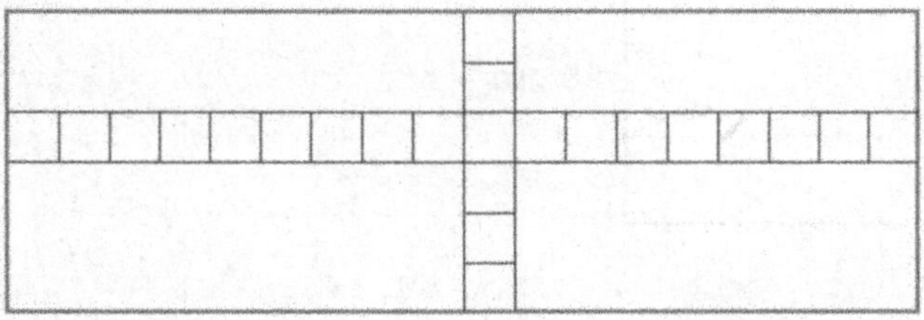

Calculating areas

Calculate the area of each of these rectangles. The rectangles are not drawn to scale.

1.

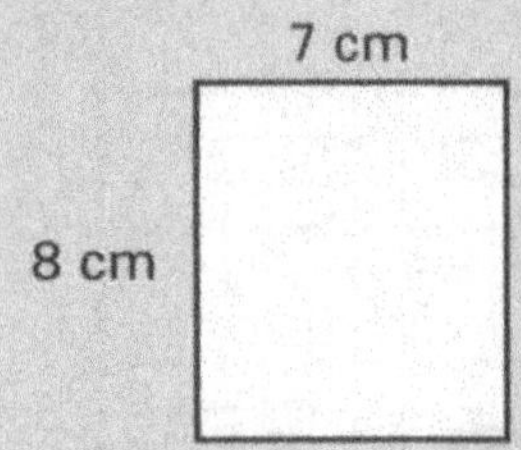

2.

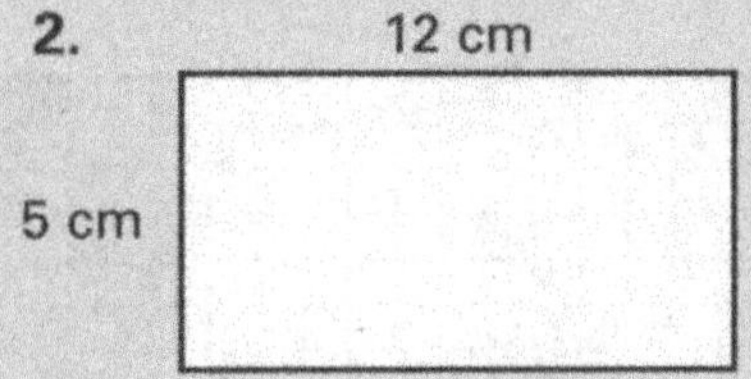

3.

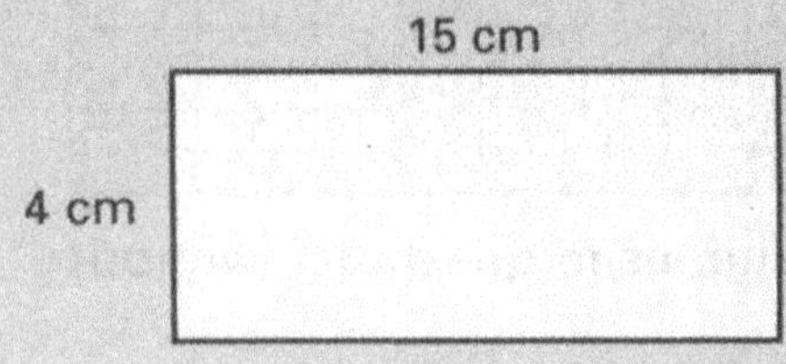

4.

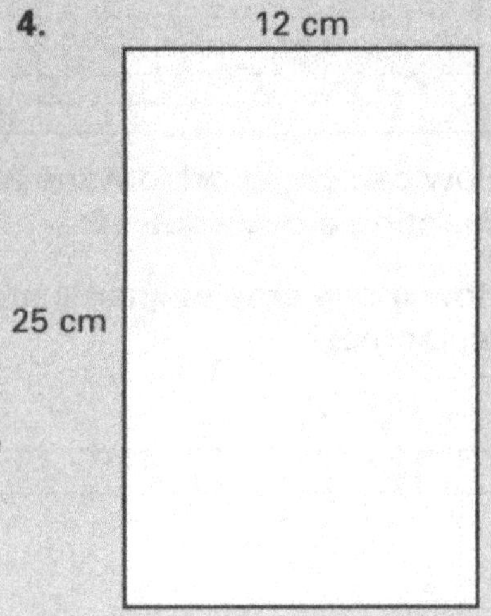

5.

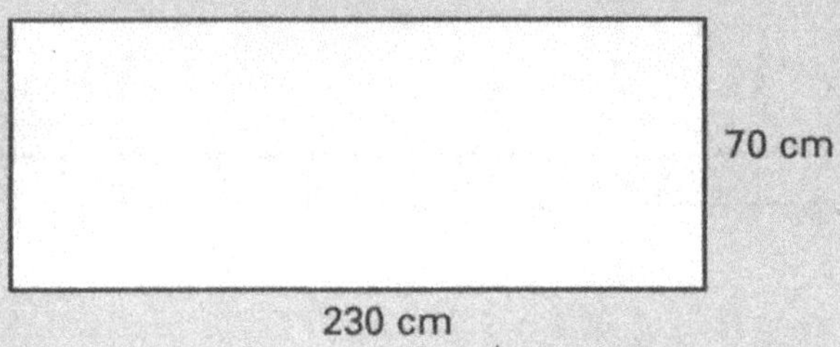

6.

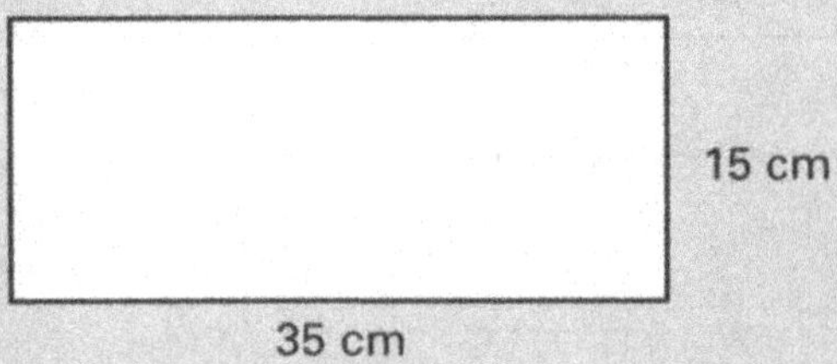

7.

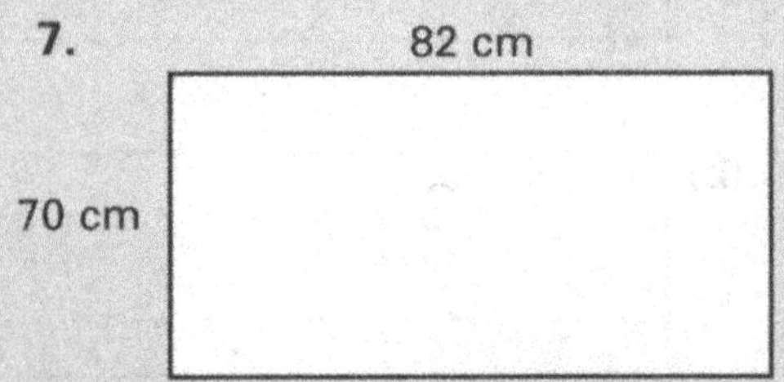

8.

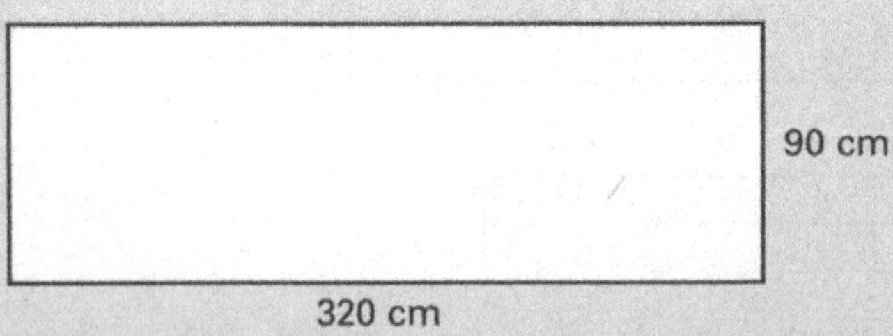

9. What is the most important thing to think about when calculating areas?

Area problems

Do the following activities on **1 cm grid paper**. This can be done by copying the 1 cm grid paper found on the inside back cover of *Student Book 5A* into your book.

1. Draw a rectangle with an area of 24 square centimetres.
2. Draw a shape that is not a rectangle with an area of 10 square centimetres.
3. Draw a triangle with an area of 6 square centimetres.
4. Draw as many different shapes as you can with an area of 8 square centimetres.
5. What letters can you draw on the grid paper that have an area of 6 square centimetres?

Do the following activities on **plain paper**.

6. Draw a diagram that represents a rectangle with an area of 36 square metres.
7. Draw a diagram that represents a square with an area of 36 square metres.
8. Draw a diagram that represents an 'L' shape with an area of 36 square metres.

9. ✎ Challenge activity ✎

I have a garden with a fence around it, that is in the shape of a rectangle. It is 10 m long and 15 m wide. I make a path 1 m wide inside the fence all the way around the garden. How much area is there for growing things?

10. ✎ Challenge activity ✎

I have a large sheet of card that is 100 cm x 150 cm. The class wants to make cards that are 10 cm x 7 cm. How many cards can I make?

Fractions of numbers

Copy these questions into your book, and work out the answer.

1. (a) $\frac{1}{2}$ of 16 = (b) $\frac{1}{2}$ of 24 = (c) $\frac{1}{2}$ of 36 =

   (d) $\frac{1}{2}$ of 200 = (e) $\frac{1}{2}$ of 860 = (f) $\frac{1}{2}$ of 424 =

   (g) $\frac{1}{2}$ of 728 = (h) $\frac{1}{2}$ of 2 240 =

2. (a) $\frac{1}{5}$ of 20 = (b) $\frac{1}{4}$ of 80 = (c) $\frac{1}{3}$ of 15 =

   (d) $\frac{1}{3}$ of 60 = (e) $\frac{1}{3}$ of 630 = (f) $\frac{1}{4}$ of 64 =

   (g) $\frac{1}{3}$ of 81 = (h) $\frac{1}{10}$ of 700 =

3. (a) $\frac{3}{5}$ of 20 = (b) $\frac{3}{4}$ of 20 = (c) $\frac{2}{3}$ of 30 =

   (d) $\frac{2}{3}$ of 15 = (e) $\frac{2}{5}$ of 40 = (f) $\frac{4}{5}$ of 100 =

   (g) $\frac{3}{10}$ of 40 = (h) $\frac{3}{10}$ of 100 =

4. (a) $\frac{1}{2}$ of 23 460 = (b) $\frac{1}{2}$ of 64 800 = (c) $\frac{1}{2}$ of 23 464 =

   (d) $\frac{1}{2}$ of 31 410 = (e) $\frac{1}{5}$ of 4 785 = (f) $\frac{1}{5}$ of 23 500 =

   (g) $\frac{1}{5}$ of 56 200 = (h) $\frac{1}{5}$ of 16 235 =

5. (a) $\frac{1}{2}$ of 7 = (b) $\frac{1}{2}$ of 15 = (c) $\frac{1}{2}$ of 3 =

   (d) $\frac{1}{2}$ of 8 = (e) $\frac{1}{2}$ of 21 = (f) $\frac{1}{2}$ of 9 =

   (g) $\frac{1}{2}$ of 101 = (h) $\frac{1}{2}$ of 51 =

6. Write down the most important thing to think about when calculating fractions of numbers.

Some fraction problems

Work out the answer to these questions. Draw a diagram if that helps.

1. How many seconds are there in $\frac{1}{2}$ of a minute?
2. How many mL are there in $\frac{1}{4}$ of a litre?
3. How many grams are there in $\frac{1}{2}$ of a kilogram?
4. How many grams are there in $\frac{3}{4}$ of a kilogram?
5. How many minutes are there in $\frac{3}{4}$ of an hour?
6. How many minutes are there in $\frac{1}{2}$ of a day?
7. How many seconds are there in $\frac{1}{4}$ of an hour?
8. If each of these containers is $\frac{1}{2}$ full, how much would each contain?

9. Challenge activity

What time is exactly half way between these times? You may want to use a clock to help you.

(a) 10:00 and 10:40
(b) 5:00 and 6:30
(c) 4:00 and 7:00
(d) 2:00 and 2:38
(e) 11:00 and 1:30
(f) 3:00 and 8:00
(g) 1:00 and 6:20
(h) 5:30 and 6:10

Fractions in their simplest terms

Copy these fractions into your book, and write them in their simplest terms.

e.g. $\frac{4}{6} = \frac{2}{3}$

1. (a) $\frac{5}{10}$ (b) $\frac{2}{8}$ (c) $\frac{3}{9}$ (d) $\frac{2}{4}$
   (e) $\frac{4}{8}$ (f) $\frac{6}{8}$ (g) $\frac{6}{9}$ (h) $\frac{2}{10}$

2. (a) $\frac{10}{20}$ (b) $\frac{4}{12}$ (c) $\frac{3}{12}$ (d) $\frac{5}{20}$
   (e) $\frac{4}{16}$ (f) $\frac{8}{16}$ (g) $\frac{10}{15}$ (h) $\frac{9}{18}$

3. (a) $\frac{20}{50}$ (b) $\frac{12}{36}$ (c) $\frac{20}{30}$ (d) $\frac{40}{60}$
   (e) $\frac{40}{50}$ (f) $\frac{25}{40}$ (g) $\frac{16}{20}$ (h) $\frac{30}{36}$

4. (a) $\frac{30}{100}$ (b) $\frac{25}{100}$ (c) $\frac{75}{100}$ (d) $\frac{20}{100}$
   (e) $\frac{60}{100}$ (f) $\frac{80}{100}$ (g) $\frac{90}{100}$ (h) $\frac{45}{100}$

5. (a) $\frac{300}{1\,000}$ (b) $\frac{250}{1\,000}$ (c) $\frac{500}{1\,000}$ (d) $\frac{800}{1\,000}$
   (e) $\frac{1\,000}{2\,000}$ (f) $\frac{1\,500}{2\,000}$ (g) $\frac{1\,200}{2\,000}$ (h) $\frac{7\,000}{2\,000}$

6. (a) $\frac{20}{24}$ (b) $\frac{7}{10}$ (c) $\frac{13}{15}$ (d) $\frac{33}{36}$
   (e) $\frac{20}{48}$ (f) $\frac{12}{25}$ (g) $\frac{14}{24}$ (h) $\frac{8}{15}$

7. Write down what you should think about when writing fractions in their simplest terms.

Addition with fractions

Copy these questions into your book, and answer the questions showing the steps you use. An example is given for each question.

1. Add the two fractions together. e.g. $\frac{1}{25} + \frac{3}{25} = \frac{4}{25}$

(a) $\frac{1}{5} + \frac{2}{5} =$ (b) $\frac{1}{3} + \frac{1}{3} =$ (c) $\frac{1}{9} + \frac{4}{9} =$

(d) $\frac{2}{7} + \frac{3}{7} =$ (e) $\frac{1}{5} + \frac{3}{5} =$ (f) $\frac{3}{11} + \frac{4}{11} =$

2. Do the addition, then write the answer in the simplest terms.
e.g. $\frac{1}{9} + \frac{2}{9} = \frac{3}{9} = \frac{1}{3}$

(a) $\frac{1}{12} + \frac{5}{12} =$ (b) $\frac{1}{10} + \frac{3}{10} =$ (c) $\frac{3}{20} + \frac{7}{20} =$

(d) $\frac{3}{8} + \frac{3}{8} =$ (e) $\frac{1}{6} + \frac{5}{6} =$ (f) $\frac{3}{10} + \frac{3}{10} =$

3. First make sure the fractions have the same denominators, then add them. e.g. $\frac{1}{2} + \frac{1}{4} = \frac{2}{4} + \frac{1}{4} = \frac{3}{4}$

(a) $\frac{1}{3} + \frac{1}{6} =$ (b) $\frac{1}{10} + \frac{1}{5} =$ (c) $\frac{1}{4} + \frac{1}{2} =$

(d) $\frac{1}{8} + \frac{1}{4} =$ (e) $\frac{5}{12} + \frac{1}{6} =$ (f) $\frac{1}{5} + \frac{7}{10} =$

4. First make sure the fractions have the same denominators, then add them. The whole number can be added separately.
e.g. $1\frac{1}{2} + \frac{1}{4} = 1\frac{2}{4} + \frac{1}{4} = 1\frac{3}{4}$

(a) $3\frac{1}{3} + \frac{1}{6} =$ (b) $4\frac{1}{2} + \frac{1}{4} =$ (c) $\frac{1}{6} + 4\frac{1}{3} =$

(d) $5\frac{1}{4} + \frac{1}{8} =$ (e) $3\frac{3}{10} + \frac{3}{20} =$ (f) $12\frac{2}{5} + \frac{3}{10} =$

5. First make sure the fractions have the same denominators, then add them. The whole numbers can be added separately.
e.g. $1\frac{1}{2} + 3\frac{1}{4} = 1\frac{2}{4} + 3\frac{1}{4} = 4\frac{3}{4}$

(a) $3\frac{1}{3} + 6\frac{1}{6} =$ (b) $4\frac{1}{2} + 3\frac{1}{4} =$ (c) $5\frac{1}{4} + 6\frac{3}{8} =$

(d) $7\frac{1}{5} + 8\frac{3}{5} =$ (e) $12\frac{7}{10} + 5\frac{1}{5} =$ (f) $103\frac{5}{8} + 407\frac{1}{4} =$

6. Write down the most important things to think about when you are adding fractions together.

Subtraction with fractions

Copy these questions into your book, and answer the questions showing the steps you use. An example is given for each question.

1. Do the subtraction. e.g. $\frac{13}{25} - \frac{2}{25} = \frac{11}{25}$

   (a) $\frac{3}{5} - \frac{2}{5} =$ (b) $\frac{7}{9} - \frac{4}{9} =$ (c) $\frac{11}{15} - \frac{7}{15} =$

   (d) $\frac{17}{25} - \frac{9}{25} =$ (e) $\frac{6}{7} - \frac{3}{7} =$ (f) $\frac{7}{11} - \frac{4}{11} =$

2. Do the subtraction, then write the answer in the simplest terms:
   e.g. $\frac{5}{9} - \frac{2}{9} = \frac{3}{9} = \frac{1}{3}$

   (a) $\frac{11}{12} - \frac{5}{12} =$ (b) $\frac{7}{10} - \frac{3}{10} =$ (c) $\frac{5}{12} - \frac{1}{12} =$

   (d) $\frac{7}{8} - \frac{3}{8} =$ (e) $\frac{5}{6} - \frac{1}{6} =$ (f) $\frac{7}{20} - \frac{3}{20} =$

3. First make sure the fractions have the same denominators, then subtract them. e.g. $\frac{1}{2} - \frac{1}{4} = \frac{2}{4} - \frac{1}{4} = \frac{1}{4}$

   (a) $\frac{1}{3} - \frac{1}{6} =$ (b) $\frac{1}{5} - \frac{1}{10} =$ (c) $\frac{1}{4} - \frac{1}{8} =$

   (d) $\frac{3}{4} - \frac{1}{2} =$ (e) $\frac{7}{10} - \frac{1}{5} =$ (f) $\frac{5}{12} - \frac{1}{6} =$

4. First make sure the fractions have the same denominators, then subtract them. The whole number can be subtracted separately.
   e.g. $1\frac{1}{2} - \frac{1}{4} = 1\frac{2}{4} - \frac{1}{4} = 1\frac{1}{4}$

   (a) $3\frac{1}{3} - \frac{1}{6} =$ (b) $4\frac{7}{10} - \frac{1}{5} =$ (c) $7\frac{1}{4} - \frac{1}{8} =$

   (d) $12\frac{2}{3} - \frac{1}{6} =$ (e) $2\frac{9}{10} - \frac{2}{5} =$ (f) $3\frac{5}{8} - \frac{1}{4} =$

5. First make sure the fractions have the same denominators, then subtract them. The whole numbers can be subtracted separately.
   e.g. $3\frac{1}{2} - 1\frac{1}{4} = 3\frac{2}{4} - 1\frac{1}{4} = 2\frac{1}{4}$

   (a) $3\frac{2}{3} - 1\frac{1}{6} =$ (b) $4\frac{1}{2} - 3\frac{1}{4} =$ (c) $5\frac{4}{5} - 1\frac{3}{10} =$

   (d) $15\frac{1}{2} - 12\frac{1}{4} =$ (e) $12\frac{7}{10} - 5\frac{1}{5} =$ (f) $57\frac{3}{8} - 47\frac{1}{4} =$

6. Write down the most important things to think about when you are subtracting fractions.

Shortcuts

A Copy these sums into your book. Work out quick ways to solve them in your head.

1. 199 + 86 = | **2.** 435 − 99 =

3. 562 − 299 = | **4.** 3 457 + 999 =

5. 989 + 101 = | **6.** 2 560 − 999 =

7. 56 − 101 = | **8.** 1 483 + 99 =

9. 399 + 1 299 = | **10.** 6 740 − 1 001 =

B Copy these number chains into your book and work out how each new number has been obtained. Write it between each number as shown in the first one.

1. 157 —(+99)— 256 —(+10)— 266 — 167 — 268 — 279

2. 507 — 407 — 508 — 519 — 420 — 1 419

3. 795 — 896 — 885 — 894 — 995 — 1 094

4. 1 563 — 563 — 1 562 — 1 573 — 2 572 — 2 473

5. 2 371 — 3 371 — 3 471 — 3 482 — 3 472 — 3 463

6. 1 608 — 1 618 — 1 718 — 1 819 — 1 810 — 1 799

7. 3 525 — 3 561 — 3 616 — 3 715 — 3 705 — 3 694

8. 631 — 1 631 — 1 620 — 1 719 — 1 708 — 1 809

Investigating 10, 100, and 1 000

A

67	5 400	365		7 620
9	410	250		4 752
630	56 000	73 000		71
75	892	4		56
3 500		126		9 045

1. Which of these numbers will have 4 digits when you multiply by 10?
2. Which numbers will have 3 digits when you divide by 100?
3. Which numbers will have a remainder when you divide by 10?
4. Which numbers will have 5 digits when you multiply by 1 000?

B

1. Write five numbers that end in 7 when they are divided by 10.
2. What could you add to 581 so that you can divide it by 10? Find as many ways as you can.
3. Write five numbers that end in 3 when they are divided by 100.
4. Write some numbers that have a remainder of 2 when they are divided by 10.
5. Write some numbers that when multiplied by 100 will have 5 digits.

More multiplying by multiples of 10

Copy these sums into your book and solve them.

A

1 (a) $\begin{array}{r} 23 \\ \times\ 10 \\ \hline \\ \hline \end{array}$ **(b)** $\begin{array}{r} 23 \\ \times\ 20 \\ \hline \\ \hline \end{array}$ **2 (a)** $\begin{array}{r} 47 \\ \times\ 10 \\ \hline \\ \hline \end{array}$ **(b)** $\begin{array}{r} 47 \\ \times\ 20 \\ \hline \\ \hline \end{array}$

3 (a) $\begin{array}{r} 35 \\ \times\ 10 \\ \hline \\ \hline \end{array}$ **(b)** $\begin{array}{r} 35 \\ \times\ 20 \\ \hline \\ \hline \end{array}$ **4 (a)** $\begin{array}{r} 76 \\ \times\ 10 \\ \hline \\ \hline \end{array}$ **(b)** $\begin{array}{r} 76 \\ \times\ 20 \\ \hline \\ \hline \end{array}$

5 (a) $\begin{array}{r} 98 \\ \times\ 10 \\ \hline \\ \hline \end{array}$ **(b)** $\begin{array}{r} 98 \\ \times\ 20 \\ \hline \\ \hline \end{array}$ **6 (a)** $\begin{array}{r} 54 \\ \times\ 10 \\ \hline \\ \hline \end{array}$ **(b)** $\begin{array}{r} 54 \\ \times\ 20 \\ \hline \\ \hline \end{array}$

7 (a) $\begin{array}{r} 69 \\ \times\ 10 \\ \hline \\ \hline \end{array}$ **(b)** $\begin{array}{r} 69 \\ \times\ 20 \\ \hline \\ \hline \end{array}$ **8 (a)** $\begin{array}{r} 42 \\ \times\ 10 \\ \hline \\ \hline \end{array}$ **(b)** $\begin{array}{r} 42 \\ \times\ 20 \\ \hline \\ \hline \end{array}$

9 (a) $\begin{array}{r} 81 \\ \times\ 10 \\ \hline \\ \hline \end{array}$ **(b)** $\begin{array}{r} 81 \\ \times\ 20 \\ \hline \\ \hline \end{array}$ **10 (a)** $\begin{array}{r} 38 \\ \times\ 10 \\ \hline \\ \hline \end{array}$ **(b)** $\begin{array}{r} 38 \\ \times\ 20 \\ \hline \\ \hline \end{array}$

B

1. 4 x 6 =
4 x 60 =
40 x 6 =

2. 3 x 8 =
3 x 80 =
30 x 8 =

3. 5 x 7 =
5 x 70 =
50 x 7 =

4. 6 x 8 =
6 x 80 =
60 x 8 =

5. 2 x 9 =
2 x 90 =
20 x 9 =

6. 7 x 4 =
7 x 40 =
70 x 4 =

Near doubles and near halves

Use doubling and halving to help you work these out.

1. 69 + 69 =
2. 231 + 231 =
3. 60 – 29 =
4. 89 x 2 =
5. 241 x 2 =
6. 85 + 84 =
7. 159 + 160 =
8. 300 – 151 =
9. 161 ÷ 2 =
10. 603 ÷ 2 =
11. 460 – 229 =
12. 159 x 2 =
13. 71 x 2 =
14. 238 ÷ 2 =
15. 402 ÷ 2 =
16. 141 + 142 =
17. 316 + 315 =
18. 2 x 199 =
19. 500 – 251 =
20. 2 x 249 =
21. 2 x 119
22. 405 ÷ 2 =
23. 79 + 81 =
24. 148 + 149 =
25. 598 ÷ 2 =
26. 490 + 505 =
27. 490 – 250 =
28. 998 – 500 =
29. 2 x 306 =
30. 2 x 297 =
31. 660 – 320 =
32. 660 – 360 =
33. 658 ÷ 2 =
34. 896 ÷ 2 =
35. 495 x 2 =
36. 399 x 2 =
37. 149 + 155 =
38. 163 + 147 =
39. 296 + 304 =
40. 444 + 556 =

Find my partner

Write the number pairs in each group that make the number shaded in red.

1. 100

63, 79, 32, 12, 45, 21, 68, 55, 37, 88, 33, 67

2. 500

256, 318, 273, 192, 244, 346, 408, 308, 227, 182, 154, 92

3. 1000

725, 596, 753, 124, 404, 876, 382, 247, 618, 91, 275, 909

4. 1200

850, 472, 110, 728, 911, 289, 253, 561, 350, 639, 1090, 947

5. Write as many pairs of numbers as you can that have a total of 1 500.

Write the missing number to complete each number sentence.

6. 1 900 – ◯ = 950

7. ◯ + 799 = 1 600

8. 5 420 – ◯ = 2 700

9. ◯ + 387= 1 400

Decimal clouds

This is a game for two players. Each player has his or her own markers or counters.

For each of these games copy the square grid into your book.

When it is your turn choose any two numbers from the cloud and either add or subtract them. Say out loud what you are doing. Place one of your counters on the answer. The person who gets four of their counters in a row is the winner.

Game 1

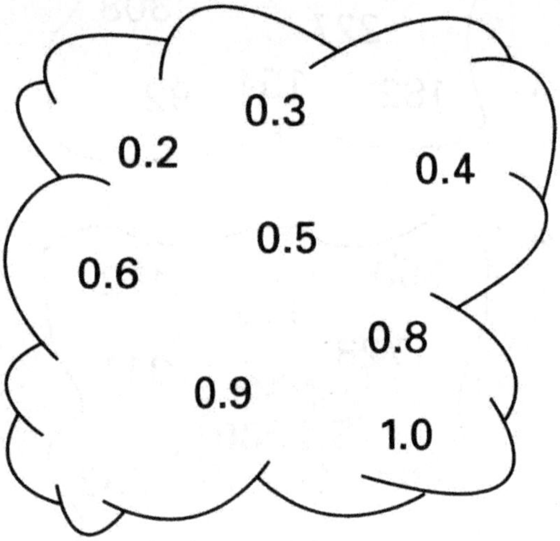

0.5	0.9	1.0	1.6
0.6	1.2	1.1	1.7
0.7	1.3	1.0	1.8
0.8	1.4	1.5	1.9

Game 2

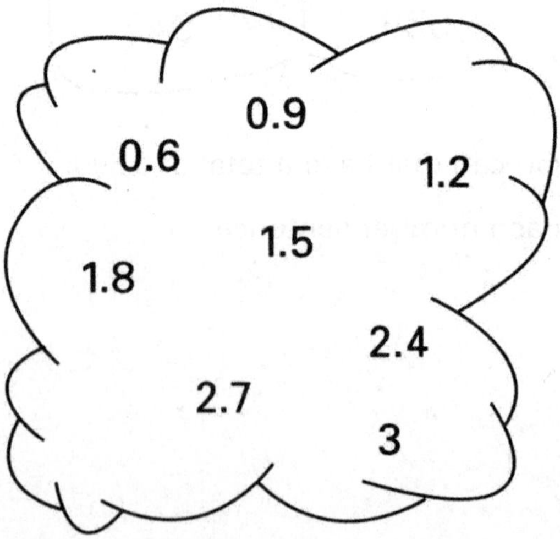

1.5	2.7	3.0	4.8
1.8	3.6	3.3	5.1
2.1	3.9	3.0	5.4
2.4	4.2	4.5	5.7

Decimal puzzles

Copy each of the diagrams into your books. Put the numbers in the boxes according to the directions.

1. 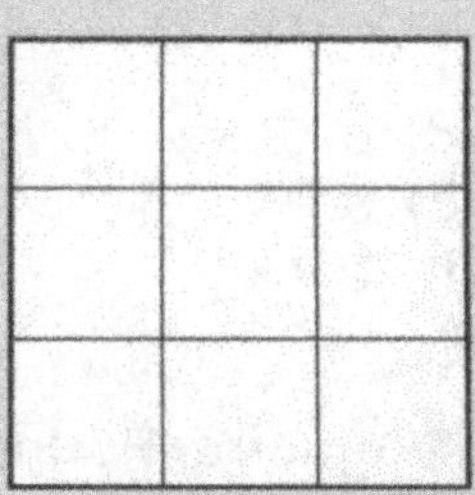

Put each of the numbers 0.1, 0.2, 0.3, 0.4, 0.5, 0.6, 0.7, 0.8, 0.9 into one of these squares so that each row and each column add to the same total.

2. 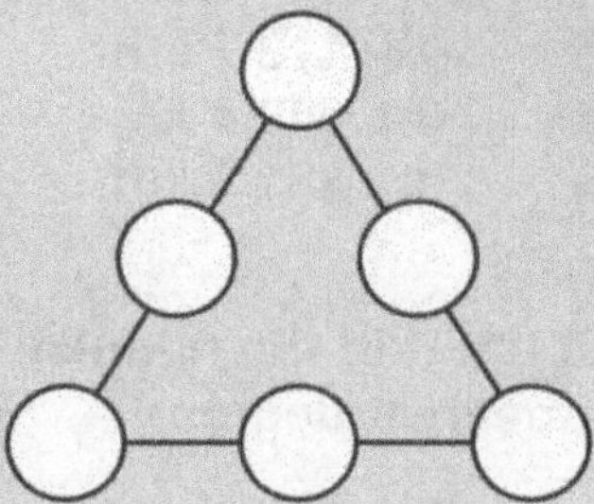

Put each of the numbers 0.2, 0.4, 0.6, 0.8, 1.0, 1.2 into one of the circles so that each line adds to the same total.

3. 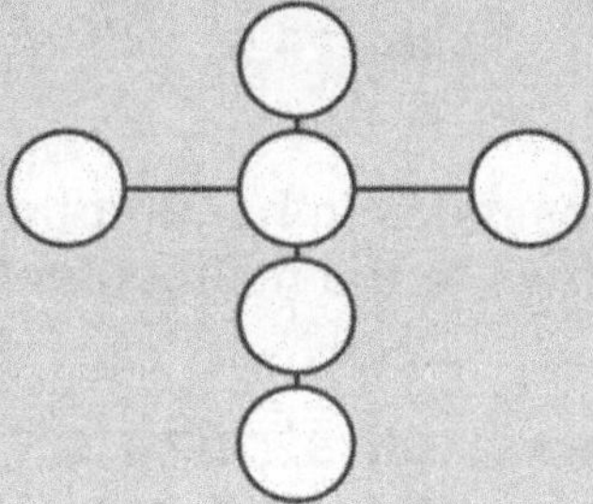

Put any of the numbers 0.5, 1.0, 1.5, 2.0, 2.5, 3.0, 3.5 into a circle so that each line adds to the same total.

4. 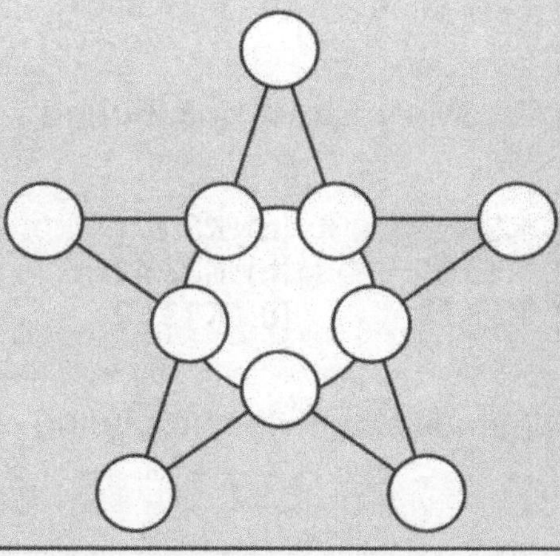

Put any of the numbers 0.1, 0.2, 0.3, 0.4, 0.5, 0.6, 0.7, 0.8, 0.9 into a circle so that each line adds to the same total.

Rounding with decimals

1. When reading a scale for weighing people, you might read the weight in tenths but we might want to round to the nearest whole number. Round these weights to the nearest whole number.

(a) 8.3 kg	(b) 28.2 kg	(c) 25.7 kg	(d) 32.8 kg
(e) 45.1 kg	(f) 36.5 kg	(g) 7.5 kg	(h) 8.2 kg
(i) 51.7 kg	(j) 13.8 kg	(k) 13.1 kg	(l) 9.9 kg

2. When giving distances we often round to the nearest kilometre. Round these distances to the nearest kilometre.

(a) 67.6 km	(b) 93.2 km	(c) 154.9 km	(d) 123.7 km
(e) 216.8 km	(f) 403.3 km	(g) 186.4 km	(h) 259.4 km
(i) 530.1 km	(j) 728.5 km	(k) 473.6 km	(l) 381.7 km

3. When giving temperatures we sometimes round them to the nearest whole degree. Round these temperatures to the nearest degree.

(a) 16.5°	(b) 27.3°	(c) 30.6°	(d) 19.8°
(e) 35.4°	(f) 37.7°	(g) 25.2°	(h) 26.9°
(i) 31.8°	(j) 39.2°	(k) 28.6°	(l) 29.5°

4. When timing runners, the stop watch sometimes reads in hundredths but we might only need tenths. Round these times to the nearest tenth of a second.

(a) 7.23 secs	(b) 15.51 secs	(c) 23.27 secs	(d) 32.38 secs
(e) 51.97 secs	(f) 32.65 secs	(g) 4.45 secs	(h) 13.35 secs
(i) 35.61 secs	(j) 32.60 secs	(k) 8.11 secs	(l) 6.27 secs

5. Sometimes money needs to be rounded to the nearest kina. Round these money amounts to the nearest kina.

(a) K3.22	(b) K5.55	(c) K367.20	(d) K8.27
(e) K12.91	(f) K13.51	(g) K55.45	(h) K32.41
(i) K32.49	(j) K91.53	(k) K85.01	(l) K13.92

6. Write down the most important thing to think about when rounding decimal numbers.

Addition with decimals 1

Answer the following questions. You could set them out like this:

$$\begin{array}{r} 24.4 \\ +\ 3_{1}1.7 \\ \hline 56.1 \\ \hline \end{array}$$

1. (a) 3.2 + 8.4 (b) 5.6 + 7.5
(c) 3.8 + 5.8 (d) 4.6 + 3.7
(e) 2.51 + 7.38 (f) 3.82 + 6.54
(g) 8.25 + 3.69 (h) 2.87 + 4.56

2. (a) 23.2 + 31.6 (b) 41.7 + 24.8
(c) 56.3 + 87.6 (d) 21.9 + 38.7
(e) 24.35 + 12.51 (f) 52.46 + 35.38
(g) 38.27 + 55.16 (h) 92.13 + 67.84

3. (a) 3.2 + 4.6 + 5.1 (b) 2.7 + 4.1 + 5.3
(c) 5.3 + 3.8 + 6.9 (d) 3.1 + 5.1 + 8.8
(e) 3. 23 + 4.61 + 5.12 (f) 4.71 + 2.63 + 3.12
(g) 1.05+ 2.81 + 5.12 (h) 3.17 + 4.92 + 3.77

4. (a) 137.2 + 158.6 (b) 251.7 + 132.1
(c) 407.1 + 522.8 (d) 315.6 + 291.7
(e) 251.32 + 181.89 (f) 256.37 + 416.92
(g) 103.12 + 200.27 (h) 257.32 + 899.97

5. (a) 2 456.3 + 1 729.4 (b) 3 617.4 + 2 912.3
(c) 6 214.7 + 2 371.4 (d) 3 020.4 + 4 907.7
(e) 5 128.32 + 6 721.44 (f) 2 060.43 + 6 921.68
(g) 5 008.07 + 3 229.87 (h) 9 997.87 + 786.58

Addition with decimals 2

Answer the following questions. You could set them out like this:

$$\begin{array}{r} 2.3 \\ +\ 7.0 \\ \hline 9.3 \\ \hline \end{array}$$

1. (a) 3.4 + 5 (b) 6 + 2.7
   (c) 4 + 1.2 (d) 3.6 + 5
   (e) 1.2 + 7 (f) 3 + 8.6
   (g) 9 + 9.2 (h) 1.3 + 1

2. (a) 7.23 + 2 (b) 5 + 3.71
   (c) 2.36 + 4 (d) 7 + 5.23
   (e) 6.77 + 5 (f) 9 + 7.23
   (g) 8 + 1.03 (h) 2.07 + 9

3. (a) 23.7 + 16 (b) 12 + 23.4
   (c) 51.8 + 14 (d) 27 + 91.2
   (e) 37.3 + 81 (f) 15 + 19.6
   (g) 10 + 13.3 (h) 20.2 + 67

4. (a) 12 + 23.45 (b) 51.26 + 20.7
   (c) 81 + 52.36 (d) 13.59 + 42
   (e) 42.35 + 17 (f) 92 + 61.73
   (g) 62.81 + 34 (h) 89 + 92.71

5. (a) 3.21 + 4.5 (b) 4.7 + 9.36
   (c) 8.25 + 7.1 (d) 7.3 + 8.05
   (e) 1.23 + 8.1 (f) 13.1 + 42.51
   (g) 7.5 + 18.05 (h) 15.67 + 1.9

6. Write down the most important thing to think about when adding decimals.

Subtraction with decimals

Answer the following questions. You could set them out like this:

$$\begin{array}{r} 2\,\overset{5}{\cancel{6}}\,.\,{}^{1}2 \\ -\;2\,4\,.\,4 \\ \hline 1\,.\,8 \\ \hline \end{array}$$

1. (a) 4.7 – 2.3 (b) 8.6 – 1.8
   (c) 7.1 – 1.4 (d) 6.2 – 3.5
   (e) 32.6 – 11.8 (f) 45.7 – 12.1
   (g) 53.2 – 27.6 (h) 81.1 – 15.9

2. (a) 2.35 – 1.28 (b) 8.27 – 5.51
   (c) 6.38 – 1.59 (d) 7.36 – 1.78
   (e) 34.81 – 11.67 (f) 53.27 – 24.61
   (g) 82.34 – 15.77 (h) 46.13 – 12.56

3. (a) 327.4 – 251.9 (b) 517.2 – 304.7
   (c) 271.8 – 148.3 (d) 806.3 – 127.8
   (e) 516.35 – 182.47 (f) 372.26 – 217.82
   (g) 912.41 – 380.29 (h) 825.81 – 118.19

4. (a) 200.3 – 15.7 (b) 50.2 – 1.8
   (c) 301.2 – 10.7 (d) 800.3 – 470.7
   (e) 5 000.0 – 1 586.3 (f) 5 060.0 – 372.1
   (g) 3 002.5 – 1 561.7 (h) 9 020.1 – 1 080.5

5. (a) 7 – 1.2 (b) 6 – 3.4
   (c) 18 – 1.3 (d) 27 – 4.7
   (e) 35 – 13.2 (f) 56 – 15.3
   (g) 100 – 1.5 (h) 80 – 33.3

6. Write down the most important thing to think about when you are subtracting with decimals.

Decimal story problems

1. One box is 1.2 m high, another box is 3.5 m, how high would the two boxes be together?
2. A large bucket contains 25.5 L of water. A small drum contains 32.8 L of water. How much water is there altogether?
3. A person travelling on a plane weighs 81.6 kg and has 22.5 kg of luggage. How heavy is that altogether?
4. On the first day of trading a store took K357.25, and on the second day they took K289.97. How much is this altogether?

5. What is the perimeter of a rectangle that is 7.8 m wide and 3.6 m long?
6. What is the perimeter of a rectangle that is 23.5 m wide and 32.7 m long?
7. A ship was carrying 2 273.3 tonnes of cargo. A further 315.8 tonnes were loaded. How much was the ship carrying altogether?

8. There are three people: one person weighs 35 kg, another weighs 42.5 kg, and another weighs 57.5 kg. How much do they weigh altogether?
9. How long is 32.5 seconds plus 16.8 seconds?
10. A piece of timber is 5.3 m long. How much more do I need to make a length of 7.1 m?

11. Tau has K31.25. Ragu has K24.80. How much more money does Tau have?
12. A water tank holds 180.5 L. If 35.8 L of water is used, how much water is left?

Recording co-ordinates

Write the corner co-ordinates of each shape.

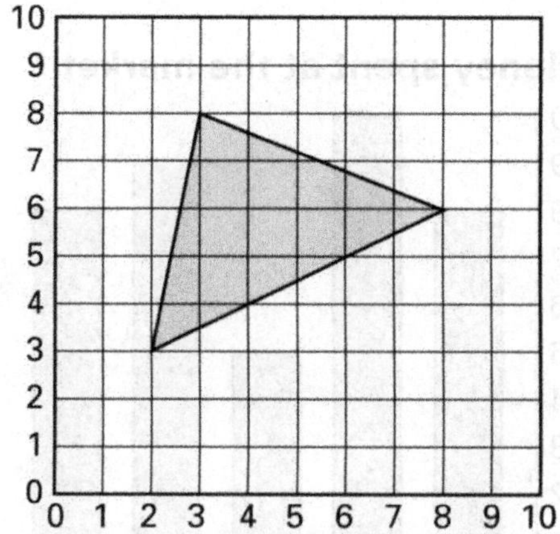

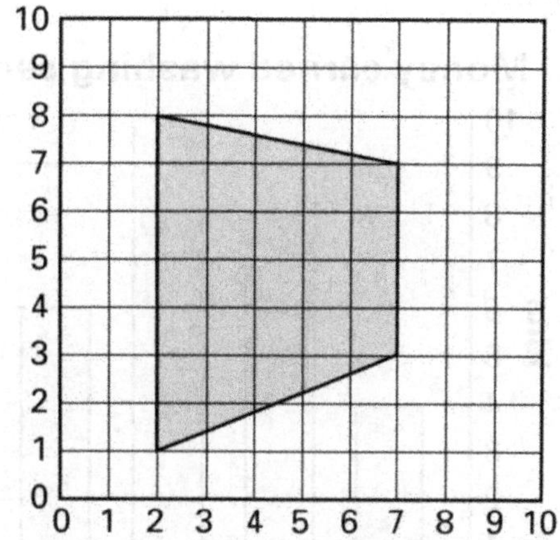

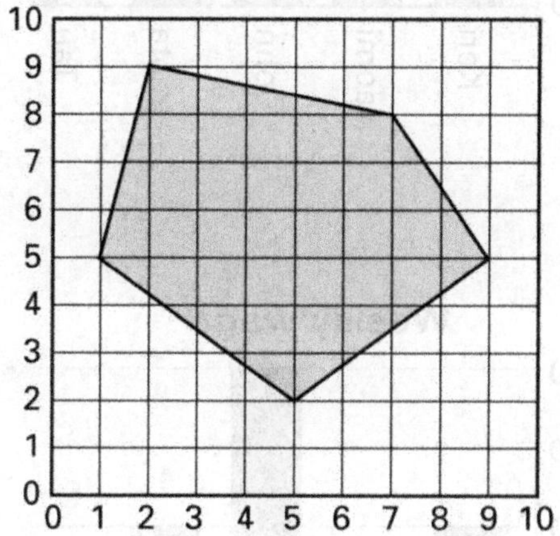

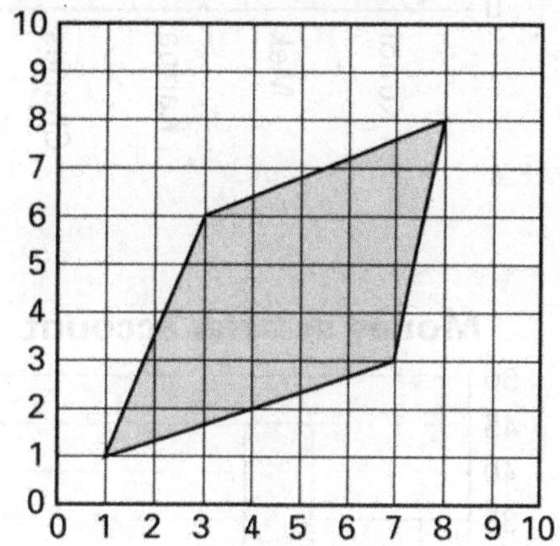

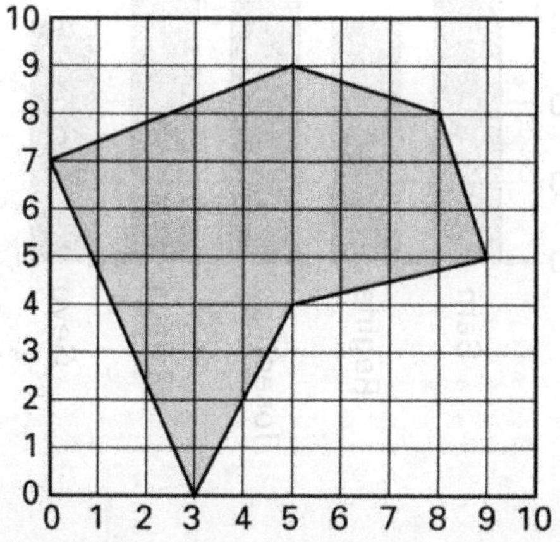

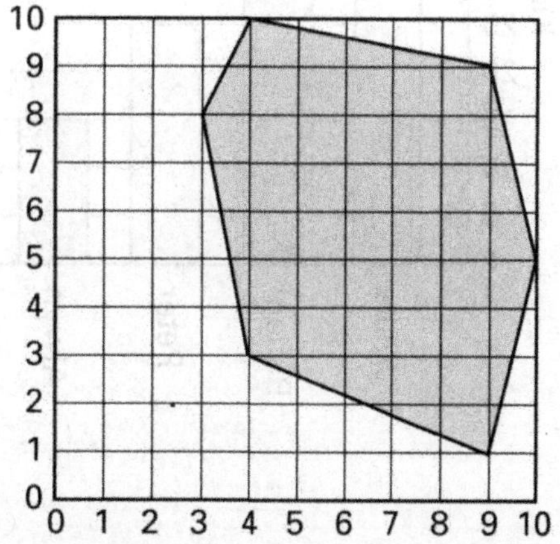

The average amount

Use coins and notes to represent each graph, then adjust them to find the mean.

1. **Money earned washing cars**

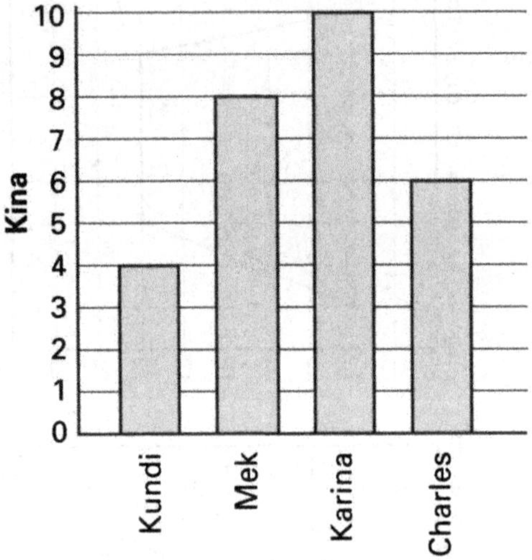

2. **Money spent at the market**

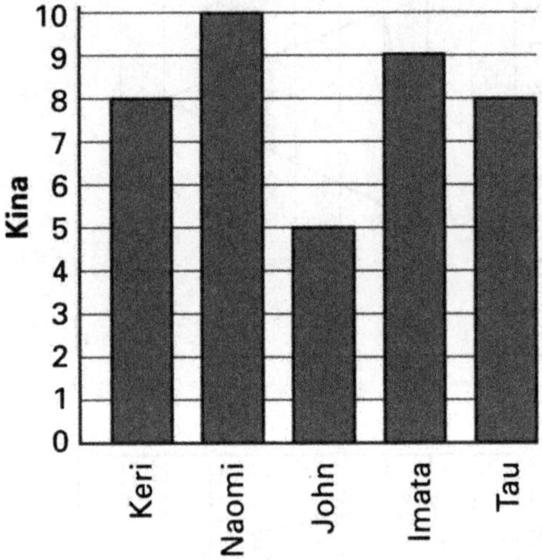

3. **Money in bank account**

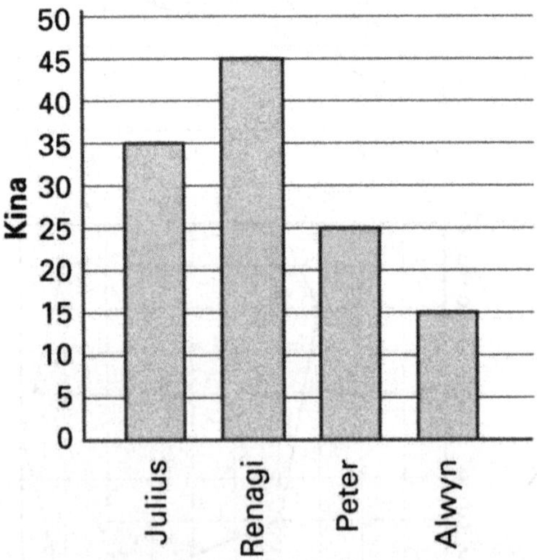

4. **Weekly wage**

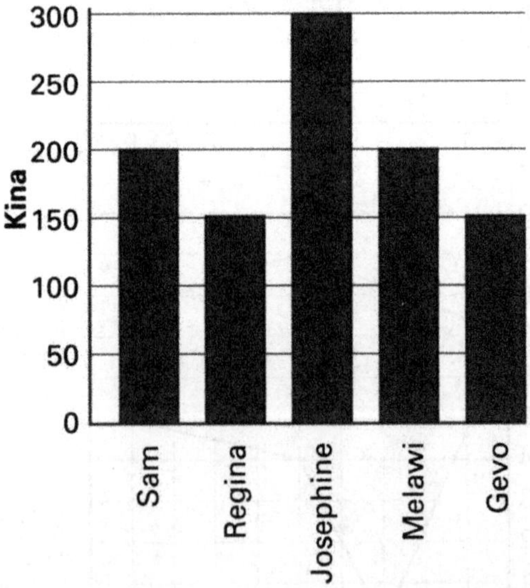

Finding the mean, mode and median 1

1. Who will make the squad? Calculate each child's average (mean) time over 100 metres and choose the four best.

Swims	Tau	Elsie	Mike	Gevo	Keri	Wep	Kundi	Julius
1st	58 sec	62 sec	57sec	60 sec	56 sec	55 sec	56 sec	65 sec
2nd	60 sec	65 sec	58 sec	60 sec	56 sec	59 sec	61 sec	66 sec
3rd	59 sec	64 sec	60 sec	58 sec	54 sec	58 sec	55 sec	64 sec
4th	63 sec	59 sec	57 sec	62 sec	58 sec	60 sec	56 sec	65 sec

2.

Temperature	Number of days	
	Tally	Frequency
16°C	II	2
19°C	IIII	4
20°C	III	3
23°C	~~IIII~~	5
24°C	~~IIII~~	5
25°C	~~IIII~~ II	7
27°C	III	3
30°C	I	1
Total		30

(a) What is the mode of this set of temperatures?
(b) What is the median?
(c) Find the mean temperature.
(d) Make a graph to show this data. Put the number of days on the vertical axis and degrees Celsius on the horizontal axis.

3. There are 5 people in a family and their average age is 28. What could their ages be?

4. The mean of three numbers is 8.5. One of the numbers is 13.9. What could the other numbers be?

Finding the mean, mode and median 2

1. Find the mean, median and mode for the set of data represented on each graph.

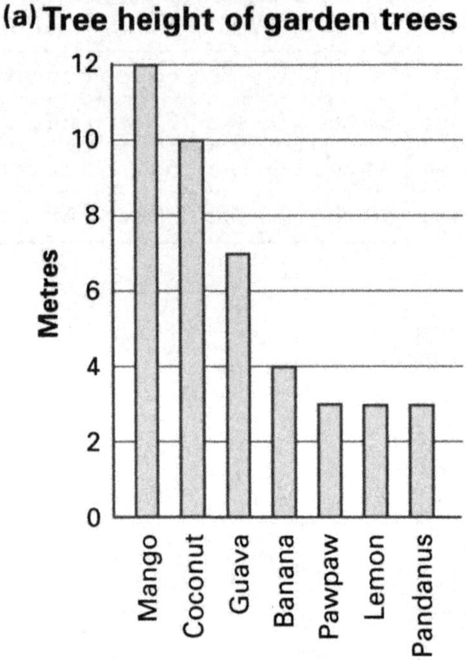

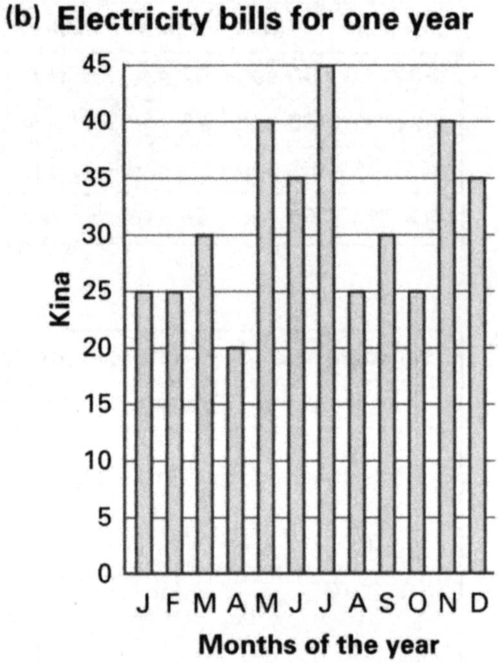

Draw the graphs in your book and rule lines across each one to show the mean, median and mode.

2. Fortnightly wages earned by workers in a cannery are: K100, K100, K100, K100, K100, K100, K100, K150, K150, K240, K300, K220. Find the mode, median and mean wages for the cannery workers.

3. Investigations

   (a) Measure the walking pace of each child in your class (the distance of one step) and find the average pace.

   (b) Collect some data about your area or school, e.g. rainfall, temperature, electricity bills etc. and find the mean, median and mode. Make a graph to show your data.

   (c) Write twenty mental arithmetic questions. Give them to the students in your class and ask them to record the answer to each question. Correct them and write the number correct out of 20 for each student. Work out the mean, median and mode for the class.

What is it about?

Write your own story to describe each graph.

1.

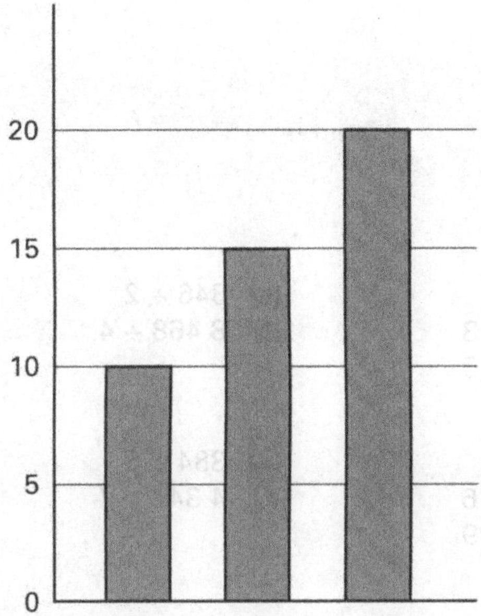

2.

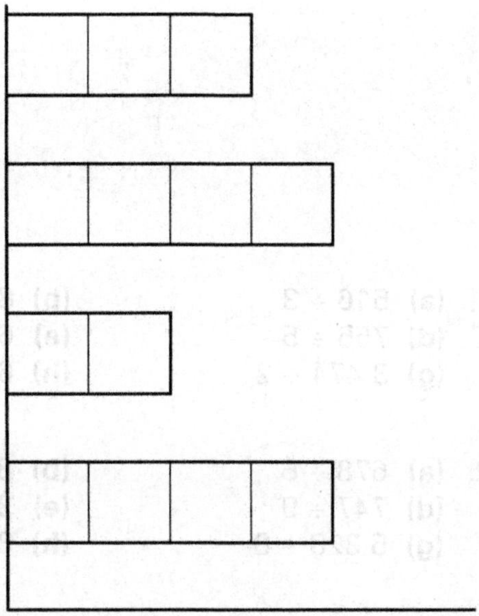

3.

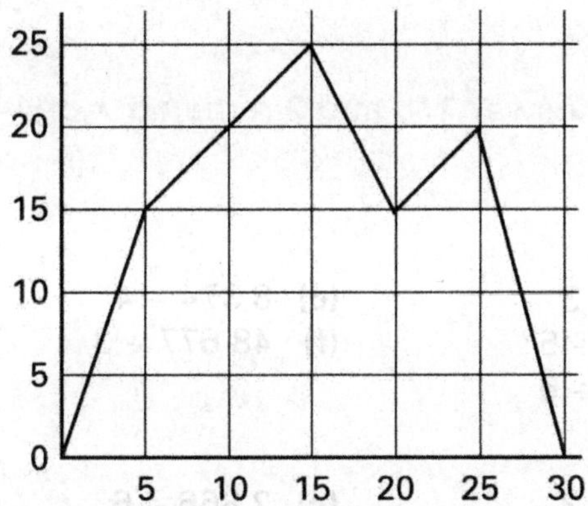

4.

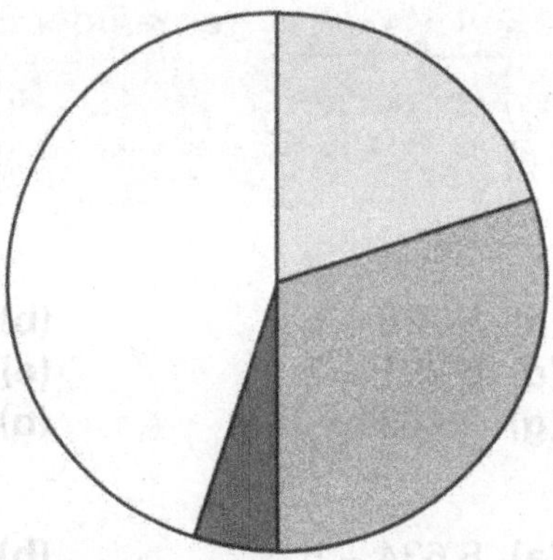

Copy each graph into your exercise book and give them titles and labels.

Division practice

A Answer these questions in your books. You could set them out like this:

$$\begin{array}{r} 1\;\;5\;\;7 \\ 3\overline{)\,4\;{}^{1}7\;{}^{2}1} \end{array}$$

Answer 157

1. **(a)** 516 ÷ 3 **(b)** 872 ÷ 4 **(c)** 346 ÷ 2
(d) 755 ÷ 5 **(e)** 6 459 ÷ 3 **(f)** 8 468 ÷ 4
(g) 3 474 ÷ 2 **(h)** 6 580 ÷ 5

2. **(a)** 678 ÷ 6 **(b)** 854 ÷ 7 **(c)** 384 ÷ 8
(d) 747 ÷ 9 **(e)** 3 726 ÷ 6 **(f)** 4 347 ÷ 7
(g) 5 328 ÷ 8 **(h)** 3 717 ÷ 9

B Answer these questions in your book. You could set them out like this.

$$\begin{array}{r} 1\;\;4\;\;8 \\ 3\overline{)\,4\;{}^{1}4\;{}^{2}6} \end{array} \quad \text{remainder 2}$$

Answer 148 and 2 remainder or $148\frac{2}{3}$

3. **(a)** 3 726 ÷ 5 **(b)** 4 867 ÷ 3 **(c)** 8 374 ÷ 4
(d) 6 291 ÷ 7 **(e)** 24 763 ÷ 5 **(f)** 48 677 ÷ 3
(g) 28 691 ÷ 8 **(h)** 96 284 ÷ 6

4. **(a)** 8 634 ÷ 5 **(b)** 8 274 ÷ 3 **(c)** 2 866 ÷ 6
(d) 3 162 ÷ 9 **(e)** 38 672 ÷ 5 **(f)** 93 811 ÷ 7
(g) 36 258 ÷ 3 **(h)** 48 229 ÷ 4

5. Write down the most important thing to think about when answering division questions.

More division

Answer these questions in your book. You could set them out like this:

```
     1 5 0 0 0   remainder 2
4 ) 6 ²0 0 0 2
```

Answer 15 000 remainder 2

or $15\,000\frac{2}{4}$

or $15\,000\frac{1}{2}$

1. (a) 4 006 ÷ 2 (b) 3 008 ÷ 4 (c) 6 060 ÷ 5
   (d) 2 004 ÷ 6 (e) 13 000 ÷ 4 (f) 26 020 ÷ 5
   (g) 57 200 ÷ 6 (h) 34 000 ÷ 8

2. (a) 3 200 ÷ 3 (b) 5 400 ÷ 7 (c) 3 006 ÷ 4
   (d) 2 020 ÷ 6 (e) 20 005 ÷ 4 (f) 30 600 ÷ 7
   (g) 60 020 ÷ 8 (h) 50 800 ÷ 3

3. (a) 624 ÷ 3 (b) 915 ÷ 5 (c) 824 ÷ 4
   (d) 912 ÷ 3 (e) 3 624 ÷ 6 (f) 2 535 ÷ 5
   (g) 8 124 ÷ 4 (h) 4 949 ÷ 7

4. (a) 9 143 ÷ 3 (b) 5 237 ÷ 5 (c) 6 919 ÷ 3
   (d) 8 137 ÷ 2 (e) 91 387 ÷ 3 (f) 82 341 ÷ 4
   (g) 62 517 ÷ 5 (h) 86 218 ÷ 8

5. (a) 6 507 ÷ 4 (b) 5 730 ÷ 7 (c) 4 163 ÷ 4
   (d) 9 665 ÷ 8 (e) 8 056 ÷ 9 (f) 10 039 ÷ 5
   (g) 12 358 ÷ 6 (h) 47 163 ÷ 4

6. Write down the most important thing to think about when answering division questions.

Long division

Answer these questions in your books. You could set them out like this:

```
          6 3
14 ) 8 8 7
   - 8 4
     ---
       4 7
     - 4 2
       ---
         5
```

Answer 63 remainder 5

1. **(a)** 432 ÷ 12 **(b)** 855 ÷ 15 **(c)** 630 ÷ 14
   (d) 546 ÷ 13 **(e)** 756 ÷ 18 **(f)** 495 ÷ 11
   (g) 288 ÷ 12 **(h)** 795 ÷ 15

2. **(a)** 437 ÷ 12 **(b)** 286 ÷ 15 **(c)** 816 ÷ 13
   (d) 392 ÷ 12 **(e)** 816 ÷ 13 **(f)** 824 ÷ 15
   (g) 413 ÷ 14 **(h)** 745 ÷ 18

3. **(a)** 3 271 ÷ 12 **(b)** 2 986 ÷ 15 **(c)** 8 673 ÷ 14
   (d) 9 216 ÷ 13 **(e)** 3 617 ÷ 18 **(f)** 8 214 ÷ 15
   (g) 3 670 ÷ 11 **(h)** 2 349 ÷ 12

4. **(a)** 14 867 ÷ 12 **(b)** 36 296 ÷ 13 **(c)** 42 386 ÷ 14
   (d) 92 167 ÷ 15 **(e)** 49 672 ÷ 16 **(f)** 38 020 ÷ 14
   (g) 20 300 ÷ 13 **(h)** 46 709 ÷ 11

5. Write down the most important thing to think about when answering these types of division questions.

Open ended division

Each of these questions has more than one possible correct answer. Give as many correct answers as you can for each question.

1. The answer to a division question is 45. What could the question be?

2. The answer to a division question is 24 with 1 remainder. What could the question be?

3. The answer to a division question is 204 and 1 remainder. What could the question be?

4. What numbers can be added to 295 so that it can be divided by 12?

5. Use the numbers 1, 2, 3 and 4 set out like □□□ ÷ □. How many different answers can you get?

6. Explain how you could calculate 23 ÷ 4 on a calculator if the '4' button is broken. Explain how you could calculate 23 ÷ 40 if the '4' button is broken?

7. In these calculations some of the numbers are missing. What could the missing numbers be?

   (a) □□ ÷ □ = 7

   (b) □□0 ÷ □ = □5

   (c) □□2 ÷ □ = □2

8. What could you add to 543 to make it divisible by 10? How many different answers can you find?

9. A number is divided by 8 and leaves a remainder of 5. What could the number be? Record as many answers as you can.

Division story problems

Write a sentence to show the answer to each of these questions.

1. A school hall has seats in rows. It can hold 408 people. There are 12 chairs in each row. How many rows are there?
2. 40 people want to travel by canoe. Each canoe can hold 12 people. How many canoes are needed?
3. 234 people want to travel by canoe. There are 28 canoes and each canoe can hold up to 8 people. How would you arrange the people?
4. 574 people want to travel by bus. Each bus can hold 20 people. How many buses are needed?

5. 5 500 people want to travel by bus. Each bus can hold 20 people. How many buses are needed?
6. 3 450 people want to travel by bus. There are 175 buses, and each bus holds a maximum of 20 people. How would you arrange the people on the buses?
7. 462 people want to travel by plane. The plane can hold 18 people. How many trips are necessary?
8. 1 210 people want to travel by plane. A plane can hold 15 people. How many planeloads are needed?
9. A recipe that makes 16 cakes includes 4 350 g of sugar. How much sugar is there in each cake?
10. A rectangle has an area of 12 570 square metres and it is 15 metres wide. How long is it?

Recording 24 hour times

1. These watches show 24 hour time. Use a.m. or p.m. to write each time.

(a)

(b)

(c)

(d)

(e)

(f)

(g)

(h)

2. Use 24 hour time to write:

(a) 7:45 a.m. **(b)** 4:56 p.m. **(c)** 2:15 a.m. **(d)** 9:30 p.m.
(e) 4:20 p.m. **(f)** 2:55 a.m. **(g)** 12:40 a.m. **(h)** 11:40 p.m.

3. What time do these planes depart? Use a.m. or p.m. to write each time.

Flight	Departure	Flight	Departure
PX 255	11:50	PX 126	16:55
PX 038	12:40	PX 046	23:25
PX 267	09:30	PX 073	22:10
PX 253	06:45	PX 145	07:40
PX 052	15:35	PX 171	17:30
PX 151	16:15	PX 158	08:50

Airline flights

Air Niugini

	MONDAY					TUESDAY					WEDNESDAY					THURSDAY					FRIDAY				
	FLT	DEP	FROM	TO	APR	FLT	DEP	FROM	TO	APR	FLT	DEP	FROM	TO	APR	FLT	DEP	FROM	TO	APR	FLT	DEP	FROM	TO	APR
F28 1000	PX154	06:05	POM	GUR	06:55																PX154	07:20	POM	GUR	06:55
	PX155	07:20	GUR	POM	08:10	PX180	07:50	POM	HGU	08:50	PX160	08:00	POM	GKA	08:50	PX180	07:50	POM	HGU	08:50	PX155	07:20	GUR	POM	08:10
	PX272	09:40	POM	RAB	11:05	PX181	09:15	HGU	POM	10:15	PX161	09:15	GKA	POM	10:05	PX181	09:15	HGU	POM	10:15	PX294	08:55	POM	MAG	09:55
		11:30	RAB	KVG	12:05	PX188	11:00	POM	HGU	12:00	PX134	11:00	POM	HGU	12:00							10:20	MAG	MAS	11:10
	PX227	12:40	KVG	MAS	13:30	PX193	12:25	HGU	MAG	12:55		12:25	HGU	WWK	13:05						PX295	11:35	MAS	MAG	12:25
		13:55	MAS	LAE	14:55		13:20	MAG	POM	14:20	PX135	13:30	WWK	HGU	14:10	PX154	14:05	POM	GUR	14:55		12:50	MAG	POM	13:50
		15:20	LAE	POM	16:05	PX154	15:20	POM	GUR	16:10		14:35	HGU	POM	15:35	PX155	15:20	GUR	POM	16:10	PX202	15:00	POM	HKN	16:05
	PX108	16:55	POM	LAE	17:40	PX155	16:35	GUR	POM	17:25	PX108	16:55	POM	LAE	17:40	PX108	16:55	POM	LAE	17:40		16:30	HKN	RAB	17:05
	PX109	18:05	LAE	POM	18:50						PX109	18:05	LAE	POM	18:50	PX109	18:05	LAE	POM	18:50	PX207	17:30	RAB	POM	18:50
	PX100	06:00	POM	LAE	06:45	PX129	07:00	WWK	MAG	07:40	PX100	06:00	POM	LAE	06:45										
	PX101	07:10	LAE	POM	07:55		08:05	MAG	LAE	08:35	PX101	07:10	LAE	POM	07:55	PX160	08:00	POM	GKA	08:50	PX160	08:00	POM	GKA	08:50
	PX160	08:40	POM	GKA	09:30		09:00	LAE	POM	09:45	PX226	08:40	POM	LAE	09:25	PX161	09:15	GKA	POM	10:05	PX161	09:15	GKA	POM	10:05
	PX161	09:55	GKA	POM	10:45	PX140	10:50	POM	WWK	12:10		09:50	LAE	MAS	10:55	PX272	11:00	POM	RAB	12:25	PX140	10:50	POM	WWK	12:10
	PX110	11:30	POM	MAG	12:30		12:35	WWK	VAI	13:15		11:20	MAS	KVG	12:10		12:50	MAS	KVG	12:10		12:35	WWK	VAI	13:15
	PX111	12:55	MAG	POM	13:55	PX141	13:40	VAI	WWK	14:20	PX273	12:45	KVG	RAB	13:20	PX223	14:00	KVG	MAS	14:50	PX141	13:40	VAI	WWK	14:20
	PX128	14.45	POM	LAE	15:30		14:45	WWK	POM	16:05		13:45	RAB	POM	15:05		15:15	MAS	MAG	16:05		14:45	WWK	POM	16:05
		15:55	LAE	MAG	16:30	PX108	16:55	POM	LAE	17:40	PX182	16:00	POM	HGU	17:00		16:30	MAG	POM	17:30	PX108	16:55	POM	LAE	17:40
		16:55	MAG	WWK	17:35	PX109	18:05	LAE	POM	18:50	PX183	17:25	HGU	POM	18:25						PX109	18:05	LAE	POM	18:50
DASH 8	PX852	06:05	POM	PNP	06:40	PX811	05:30	LNV	POM	07:35	PX811	05:30	LNV	POM	07:35	PX852	06:05	POM	PNP	06:40	PX811	05:30	LNV	POM	07:35
	PX853	07:00	PNP	POM	07:35	PX894	08:30	POM	WBM	09:55	PX866	08:30	POM	TIZ	10:00	PX853	07:00	PNP	POM	07:35	PX860	08:30	POM	MDU	09:55
	PX860	08:15	POM	MDU	09:40		10:15	WBM	MDU	10:40		10:20	TIZ	MDU	10:40	PX894	08:30	POM	WBM	09:55	PX861	10:15	MDU	POM	11:35
	PX861	10:00	MDU	POM	11:20	PX861	11:00	MDU	POM	12:20	PX861	11:00	MDU	POM	12:20		10:15	WBM	MDU	10:40	PX862	12:15	POM	CMU	13:30
	PX862	12:00	POM	CMU	13:15						PX862	13:00	POM	CMU	14:15	PX861	11:00	MDU	POM	12:20	PX863	13:50	CMU	POM	15:00
	PX863	13:35	CMU	POM	14:45						PX863	14:35	CMU	POM	15:45						PX858	15:40	POM	GUR	16:40
	PX810	15:30	POM	LNV	17:40	PX810	15:30	POM	LNV	17:40	PX856	16:30	POM	PNP	17:05						PX859	17:00	GUR	POM	18:00
											PX857	17:25	PNP	POM	18:00	PX810	15:30	POM	LNV	17:40					
						PX852	06:05	POM	PNP	06:40	PX852	06:05	POM	PNP	06:40						PX852	06:05	POM	PNP	06:40
	PX864	09:55	POM	TIZ	11:35	PX853	07:00	PNP	POM	07:35	PX853	07:00	PNP	POM	07:35	PX811	07:30	LNV	POM	09:40	PX853	07:00	PNP	POM	07:35
	PX865	11:55	TIZ	POM	13:35	PX870	08:15	POM	GKA	09:25											PX864	08:15	POM	TIZ	09:55
						PX871	09:45	GKA	POM	10:55											PX865	10:15	TIZ	POM	11:55
											PX818	11:20	POM	MAG	12:45						PX888	12:35	POM	HGU	14:00
											PX819	13:05	MAG	POM	14:30						PX889	14:20	HGU	POM	15:45
	PX856	16:30	POM	PNP	17:05	PX856	16:30	POM	PNP	17:05	PX810	15:30	POM	LNV	17:45	PX856	16:30	POM	PNP	17:05	PX856	16:30	POM	PNP	17:05
	PX857	17:25	PNP	POM	18:00	PX857	17:25	PNP	POM	18:00						PX857	17:25	PNP	POM	18:00	PX857	17:25	PNP	POM	18:00

POM = Port Moresby
GUR = Alotau
RAB = Rabaul
KVG = Kavieng
MAS = Manus Is.
HGU = Mt. Hagen
MDU = Mendi
LNV = Lihir
TIZ = Tari
MAG = Madang
GKA = Goroka
WWK = Wewak
HKN = Hoskins
VAI = Vanimo
PNP = Popondetta
CMU = Kundiawa
WBM = Wapenamanda

Look at the timetable in pairs and ask three questions about it.
Write them on a separate piece of paper.

Case study: Suburban trains in Melbourne

		A.M.	A.M	A.M	A.M	A.M	A.M	A.M	A.M	A.M	A.M
Box Hill	DEPART	7 02	7 06	7 11	7 19	7 23	7 26	7 32	7 37	7 40	7 45
Mont Albert		7 04	7 08	7 13	*	*	*	*	*	*	*
Survey Hills		7 05	7 09	7 14	*	*	*	*	*	*	*
Chatham		7 07	7 11	7 16	*	*	*	*	*	*	*
Canterbury		7 09	7 13	7 18	*	*	*	*	*	*	*
East Camberwell		7 11	7 15	7 20	*	*	*	*	*	*	*
Camberwell		7 13	7 17	7 22	7 26	7 30	7 33	7 39	7 44	7 47	*
Auburn		*	*	*	*	*	*	*	*	*	*
Glenferrie		7 16	7 20	*	*	7 33	7 36	*	7 47	7 50	*
Hawthorn		*	*	*	*	*	*	*	*	*	*
Burnley		*	*	*	*	*	*	*	*	*	*
East Richmond		*	*	*	*	*	*	*	*	*	*
Richmond		7 22	7 26	7 30	7 34	7 39	7 42	7 47	7 53	7 56	7 59
Parliament		7 25	7 29	7 33	7 37	7 42	7 45	7 50	7 56	7 59	8 02
Melbourne Central		7 27	7 31	7 35	7 39	7 44	7 47	7 52	7 58	8 01	8 04
Flagstaff		7 29	7 33	7 37	7 41	7 46	7 49	7 54	8 00	8 03	8 06
Spencer Street		7 31	7 35	7 39	7 43	7 48	7 51	7 56	8 02	8 05	8 08
Flinders Street	ARRIVE	7 35	7 39	7 43	7 47	7 52	7 56	8 00	8 06	8 09	8 12

		P.M.	P.M	P.M	P.M	P.M	P.M	P.M	P.M	P.M	P.M
Flinders Street	DEPART	5 16	5 27	5 22	5 24	5 29	5 41	5 33	5 39	5 45	5 57
Spencer Street		5 19		5 25	5 27	5 32		5 36	5 42	5 48	
Flagstaff		5 21	↓	5 27	5 29	5 34	↓	5 38	5 44	5 50	↓
Melbourne Central		5 23	↓	5 29	5 31	5 36	↓	5 40	5 46	5 52	↓
Parliament		5 25	↓	5 31	5 33	5 38	↓	5 42	5 48	5 54	↓
Richmond		5 28	5 30	5 34	5 36	5 41	5 44	5 45	5 51	5 57	6 00
East Richmond		*	5 32	*	*	*	5 46	*	*	*	6 02
Burnley		*	5 34	*	*	*	5 48	*	*	*	6 04
Hawthorn		*	5 36	*	*	*	5 50	*	*	*	6 06
Glenferrie		5 33	5 38	*	*	5 46	5 52	*	*	6 02	6 08
Auburn		*	5 40	*	*	*	5 54	*	*	*	6 10
Camberwell		5 37	5 42	*	5 44	5 50	5 56	5 53	5 59	6 06	6 12
East Camberwell		*	5 44	*	*	*	5 58	*	*	*	6 14
Canterbury		*	5 46	*	*	*	6 00	*	*	*	6 16
Chatham		*	5 48	*	*	*	6 02	*	*	*	6 18
Surrey Hills		*	5 50	*	*	*	6 04	*	*	*	6 20
Mont Albert		*	5 52	*	*	*	6 06	*	*	*	6 22
Box Hill	ARRIVE	5 44	5 54	5 48	5 51	5 57	6 08	6 00	6 06	6 13	6 24

* means the train does not stop at this station
↓ different route

Use this train timetable to answer these questions.

1. If Tom Bennett catches the 7:02 a.m. train at Box Hill station, what time will he arrive at Flinders Street? How long does the journey take?
2. If Tom does not catch a train until 7:45 a.m. how long will his journey to Flinders Street take?
3. If David catches a train at Melbourne Central at 5:31 p.m. what time will he arrive at Box Hill?
4. On Wednesday evening Alice and Sarah finish sports practice at 5:30 p.m. and arrive at Glenferrie station at 5:50 p.m. What is the first train they can catch? What time will they arrive at Box Hill station? If they miss this train, what time is the next one?

Solving time problems

Solve each of these problems.

1. A generator uses 50 L of water every second. How many litres pass through the generator in one minute?
2. If Naomi earns K4 200 per year, how much does she earn each month?
3. Suari walks 200 metres every two minutes. How many minutes will it take him to walk 1 kilometre? 2 kilometres? 5 kilometres?
4. The local PMV driver drove for 5 hours and 40 minutes on Friday and for 6 hours and 35 minutes on Saturday. How long did he drive for over the two days?

5. Ben can earn K120 each fortnight, for a ten day fortnight. How much is this per day? How much would he earn per week? If he works 8 hours each day, how much does he earn per hour? How much would he earn each month? (work it out according to the number of days in each month) What is his yearly wage?

6. A large plane takes 5 hours 10 minutes to reach its destination. A smaller plane takes 7 hours 50 minutes to reach the same destination. How much longer does the smaller plane take?
7. Kassage boarded a plane at 14:45. The flight took 55 minutes. At what time did the flight land? Give the time in both 24 hour time and 12 hour time.

8. Susi needs six hours to cut and sew a meri blouse. How many meri blouses can she complete in ten days if she works 10 hours every day?

More time problems

Solve these problems:

1. Feareka took her first dose of medicine at 10:30 a.m. What times will she take the next 3 doses?

2. A PMV left Lae at 11:55 a.m. and arrived at Goroka 6 hours and 20 minutes later. What time did it arrive?
3. A graduation dinner will take 2 hours 20 minutes to prepare. We want the dinner to start at 7:15 p.m. What time should we start preparing?
4. I left home at 10:50 a.m. and arrived at 7:05 p.m. How long was my journey?
5. Look at the following information then answer the questions.

Name	Arrive	Depart
Karl	6:15 a.m.	9:30 a.m.
Joe	8:30 a.m.	5:15 p.m.
Rani	7:00 a.m.	1:50 p.m.
Casi	10:45 a.m.	7:15 p.m.
Regina	9:15 a.m.	3:30 p.m.
Nounah	1:00 p.m.	6:45 p.m.

(a) How long does each person work?
(b) Who works for the longest time?
(c) How many minutes does Rani work for?
(d) Who works less than 6 hours?

Finding profit and loss

1. Copy these tables into your exercise book and complete them.

(a)

Buying price	Selling price	Profit	Loss
K125	K200		
K2 500	K2 800		
K1 550	K1 530		
K3 174	K3 026		
K5 036	K5 754		
K1 060	K975		
K1 985	K2 080		
K3 960	K4 100		

(b)

Buying price	Selling price	Profit	Loss
K3 150	K3 120		
K5 000	K4 870		
K1 930	K2 195		
K878	K1 216		
K959	K1 099		
K2 099	K1 999		
K4 250	K5 600		
K1 290	K2 520		

2. Work out answers to these problems.

(a) Kiri buys radio cassettes for K174 each and sells them for K201. What is his total profit on 6 radio cassettes?

(b) William bought three large bunches of bananas, one for K15, one for K12 and one for K13. He broke them into a total of 24 hands and sold each hand for K2.50. What was his profit?

(c) Lena bought a bag of 350 betel nuts for K60. If she sells each betel nut for 30t how much profit will she make?

(d) Mike sells soccer balls in his tradestore. He makes a profit of K8.15 every time he sells a soccer ball. If he paid K2.90 for each one, what was his selling price?

(e) Mirou bought 3 bags of onions for K47.50 each. Each bag had 87 onions, which she sold for 60 toea each. What was her profit?

How much profit?

1. Work out the profit for each item. Write the product's name and the profit in your exercise book.

(a)
Cost price K100
Selling price K150

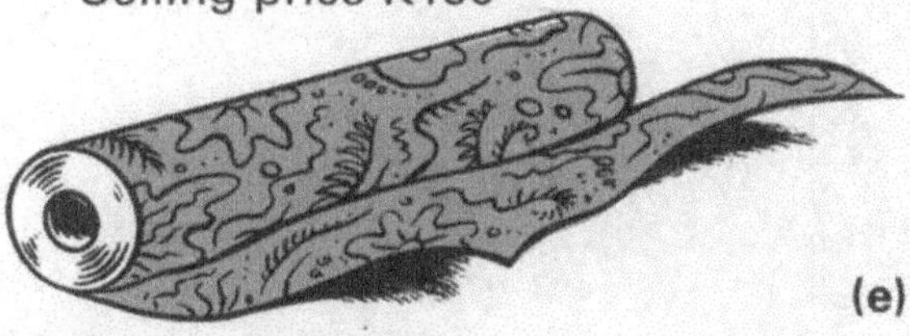

(b) Cost price K120
Selling price K154

(c)
Cost price K6.50
Selling price K10.95

(d)
Cost price K24.50
Selling price K40.95

(e)
Cost price K85
Selling price K120

(f)
Cost price K25.50
Selling price K35.95

Use the prices of the items to work out answers to these questions.

2. How much profit will the shop owner make if he sells:

(a) 2 bolts of material and 6 cartons of tinned fish?
(b) 3 buckets and 1 carton of tinned fish?
(c) 4 mosquito nets and 2 clocks?

3. (a) If the shop owner buys 10 buckets, 6 cartons of fish, 6 mosquito nets, 8 clocks, 4 mattresses and 5 bolts of material, how much will he spend?
(b) If he sells all these things how much will he make?

Case study: A monthly budget

These are some of the monthly expenses for the Maliaki family.

Items	Cost per month
Rent	K480
Water	K36
Electricity	K57
Telephone	K63
PMV fare	K40

List some other things and their cost that the family would have to pay for each month.

Answer the following questions.

1. What things will they have to pay for once a year?
2. What things in the local community might they have to pay for?
3. What other things might the family sometimes have to find money for?
4. Will the family have enough money to meet all their expenses? How could you work this out?

Case study: A yearly budget

The table shows some of the things that the Siwi family need to spend money on every year. A few have been done for you. Copy the table into your book, complete it and add any extra items you can think of.

Item	Cost
Ellen's school fees	K300
Mark's school fees	K100
Garden tools: 1 spade 1 bushknife	 K15 ....
PMV fares	....

Answer the following questions after the teacher has read the case study to you.

1. Make a list of the ways that the Siwi family raises money. Estimate how much money they could make each year from each of the things on the list.

2. When will the Siwi family need most money each year? How can they prepare for this?

3. Make a list of some unexpected costs the Siwl Family could have in a year.

4. How could the Siwi family prepare for these unexpected costs?

5. Which money-raising activity do you think the family spends most time on?

6. Which money-raising activity do you think raises the most money for the family?

Converting between volume and weight

1. What is the weight of the water in each container?

(a)

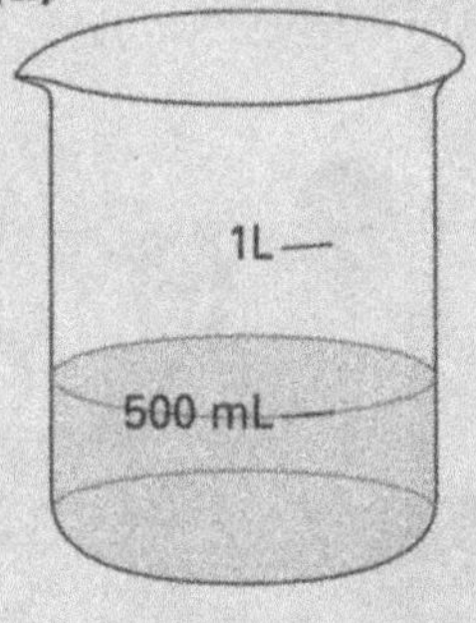

(b)

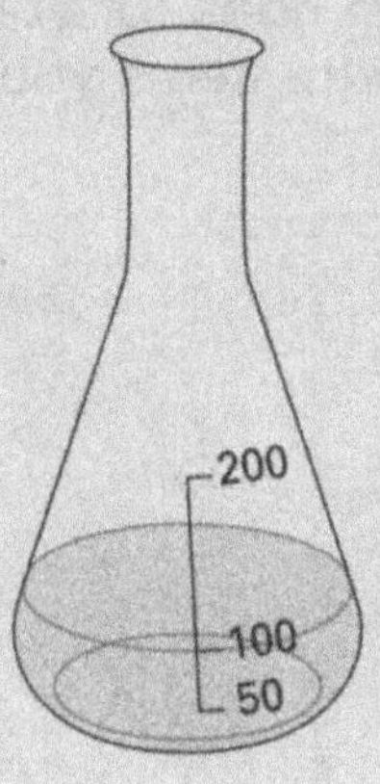

(c)

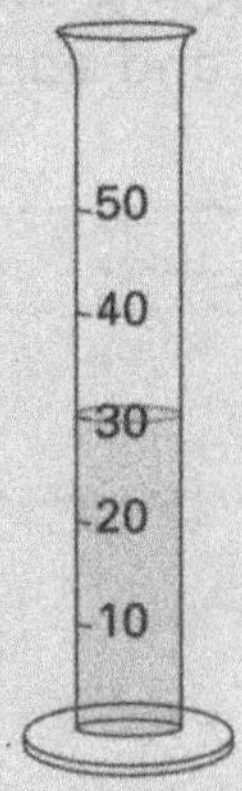

(d)

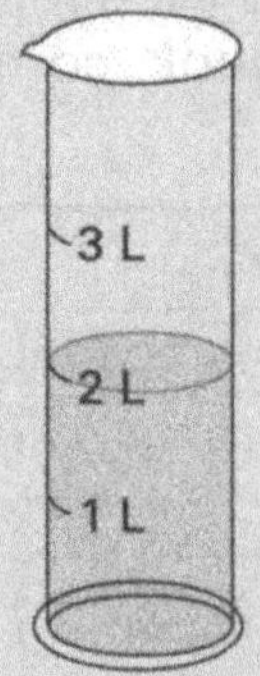

2. How much water is in the bucket on the left?

(a)

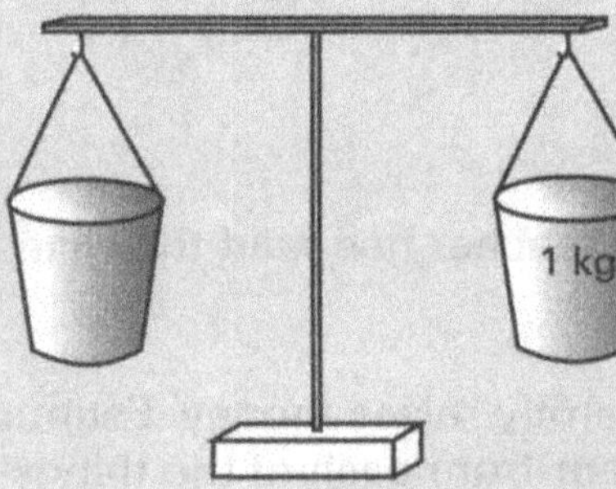

(b)

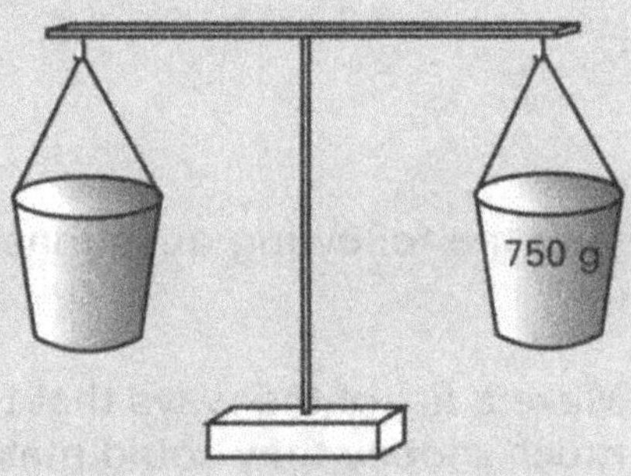

3. Write the weight of each volume of water in grams or kilograms:

(a) 10 L	**(b)** 275 mL	**(c)** 50 mL	**(d)** 750 mL
(e) 2 L	**(f)** 2.5 L	**(g)** 0.6 L	**(h)** 325 mL
(i) 4.6 L	**(j)** 480 mL	**(k)** 595 mL	**(l)** 5 L
(m) 600 mL	**(n)** 835 mL	**(o)** 3.2 L	

4. Write the volume for each weight of water in millilitres or litres:

(a) 570 g	**(b)** 0.8 kg	**(c)** 5 kg	**(d)** 75 g
(e) 215 g	**(f)** 5.6 kg	**(g)** 33 g	**(h)** 650 g
(i) 0.7 kg	**(j)** 3.5 kg	**(k)** 485 g	**(l)** 890 g
(m) 5 g	**(n)** 1.2 kg	**(o)** 4 kg	

Solving word problems

1. A water container weighs 1 530 kg when empty. If it carries 950 L what will be its total weight?

2. A container has a weight of 85 g. When it is filled with water, it has a gross weight of 2.5 kg. What volume of water does the container hold?

3. A 5 L container was filled with water. The weight of the empty container was 120 grams. How heavy was the container and its contents?

4. Eli's water flask weighed 565 g. She had put 500 mL of water in it. How heavy was the container?

5. A 200 L water tank weighs 15 kg when empty. How much does it and its contents weigh when it is full?

6. What is the weight of the water in a pool that is 10 metres long, 5 metres wide and 1 metre deep?

7. A bucket weighs 150 g. When it is filled with water, it has a total weight of 8.6 kg. What volume of water does the bucket hold?

8. How heavy will this fish tank be? It is full of water. Would you be able to pick it up?

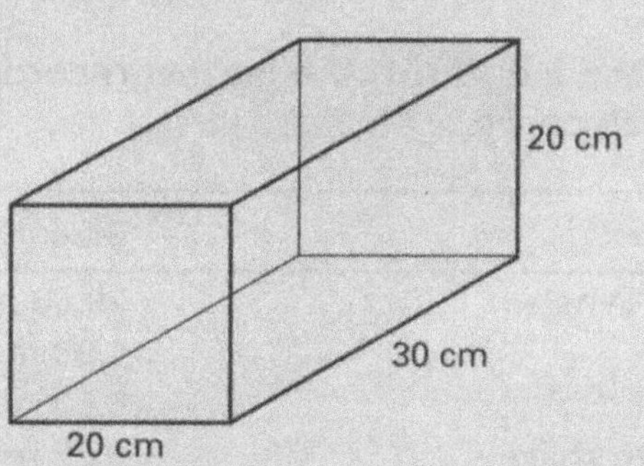

9. Ten students placed their full drink flasks into a box. Each flask contained 600 mL of water and each empty flask weighed 70 g. The box weighed 200 g. How heavy was the box of drinks?

10. A bucket holds $8\frac{1}{2}$ litres of water. The empty bucket weighs 100 g. How heavy is the bucket filled with water?

Records

1. The following are the men's and women's PNG national records in athletics events. The list includes record time, distance, record holder, and year the record was set.

Event	Men	Women
High jump	1.98 m Alfred Antipas 1993	1.7 m Nettie Elisha 1982
Long jump	7.18 m Kila Raula 1971	5.75 m Lily Tua 1991
Triple jump	15.11 m Evan Iewago 1973	11.48 m Angela Way 1990
Pole vault	4.20 m Eric Momberger 1992	none
Discus	45.62 m Albert Momberger 1990	38.46 m Iamo Launa 1991
Shotput	13.80 m Albert Momberger 1992	13.50 m Iamo Launa 1991
Javelin	58.62 m Albert Momberger 1991	51.90 m Iamo Launa 1993

Choose some of these records.

(a) Estimate and mark out on the ground how long the distance of the jump or the throw would be.

(b) Measure out the distance that the record holder jumped or threw. Compare with your estimate.

(c) See how far you can jump or throw.

2. The following are the lengths of the largest recorded animals from the *Guinness Book of Records*.

Animal	Size
Blue Whale	29.56 m long
Emu	1.93 m tall
Crocodile	8.63 m long
Whale shark	9.22 m long
Tarantula spider	24 cm across
Spider crab	2.13 m across

Estimate these lengths and mark them on the ground, then measure your estimates to see how close they were.

Writing lengths using decimals

1. Write these lengths in metres
e.g. 153 cm = 1.53 m

(a) 260 cm (b) 700 cm (c) 875 cm
(d) 175 cm (e) 2 250 cm (f) 250 cm
(g) 2 500 cm (h) 8 225 cm

2. Write these lengths in metres and centimetres
e.g. 3.20 m = 3 m 20 cm

(a) 3.25 m (b) 5.52 m (c) 7.25 m
(d) 5.99 m (e) 15.50 m (f) 25.75 m
(g) 12.05 m (h) 30.02 m

3. Write these lengths in kilometres
e.g. 3 250 m = 3.250 km

(a) 1 800 m (b) 6 200 m (c) 1 224 m
(d) 7 150 m (e) 82 250 m (f) 17 000 m
(g) 45 500 m (h) 16 600 m

4. Write these lengths in metres
e.g. 2.250 km = 2 250 m

(a) 5.500 km (b) 7.250 km (c) 8.587 km
(d) 1.250 km (e) 12 km (f) 15.500 km
(g) 12.050 km (h) 32.200 km

5. Which length in each pair is the shorter?

(a) 246 cm or 2.4 m (b) 185 cm or 1.89 cm
(c) 6.3 m or 632 cm (d) 400 cm or 3.96 cm
(e) 15.08 m or 1 500 cm (f) 6.9 m or 687 cm
(g) 1 325 m or 1.3 km (h) 2.563 km or 2 500 m
(i) 15 km or 1 550 m (j) 3.7 km or 3 675 m
(k) 5 200 m or 5.25 km (l) 2 700 m or 2.65 km

Five square shapes

1. Calculate the perimeter and area of these shaded five square shapes.

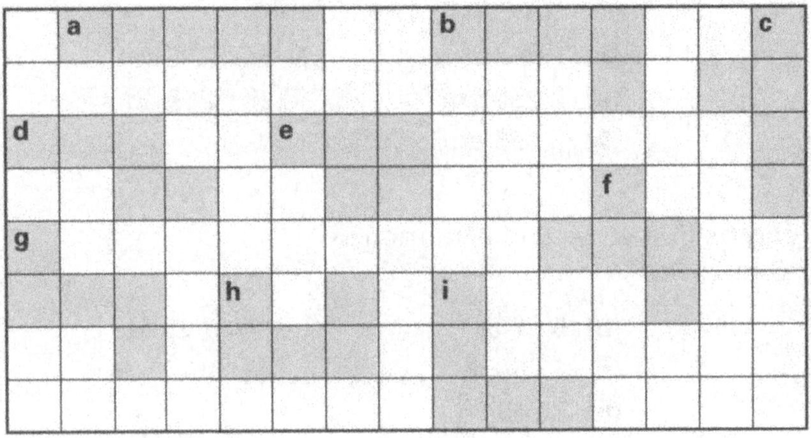

2. There are 3 more five square shapes that are different from those drawn above. Draw the other 3 five square shapes and calculate their perimeter and area. Note that to be different the shape must not be able to be turned or flipped to make one of these shapes. For example, these shapes below are all the same as shape 'b' in question 1.

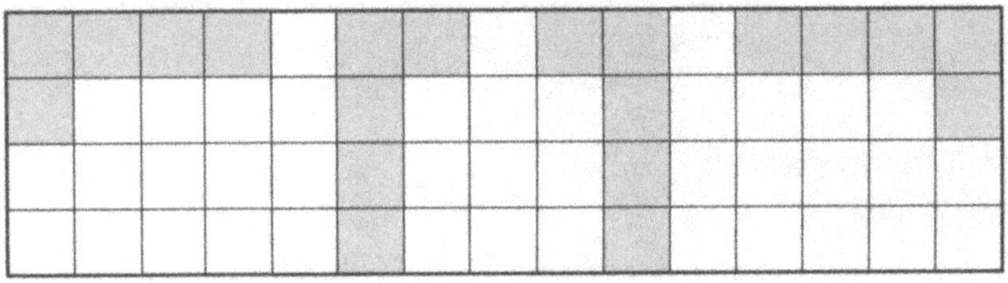

3. ✎ Challenge activity ✎

Some of these five square shapes are the nets of open top boxes. This means that if you cut them out from card, then fold along the lines, they could be folded to boxes without a top. Which ones are the nets of open top boxes? For example shape **(f)** can be folded into the box below (with 5 sides but no top)

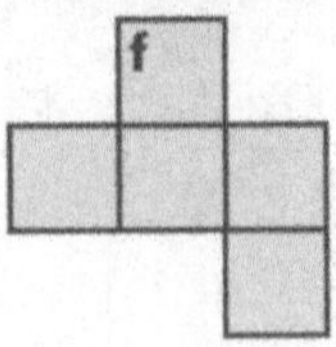

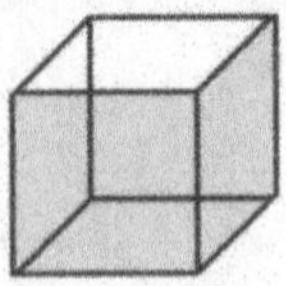

Map of Papua New Guinea

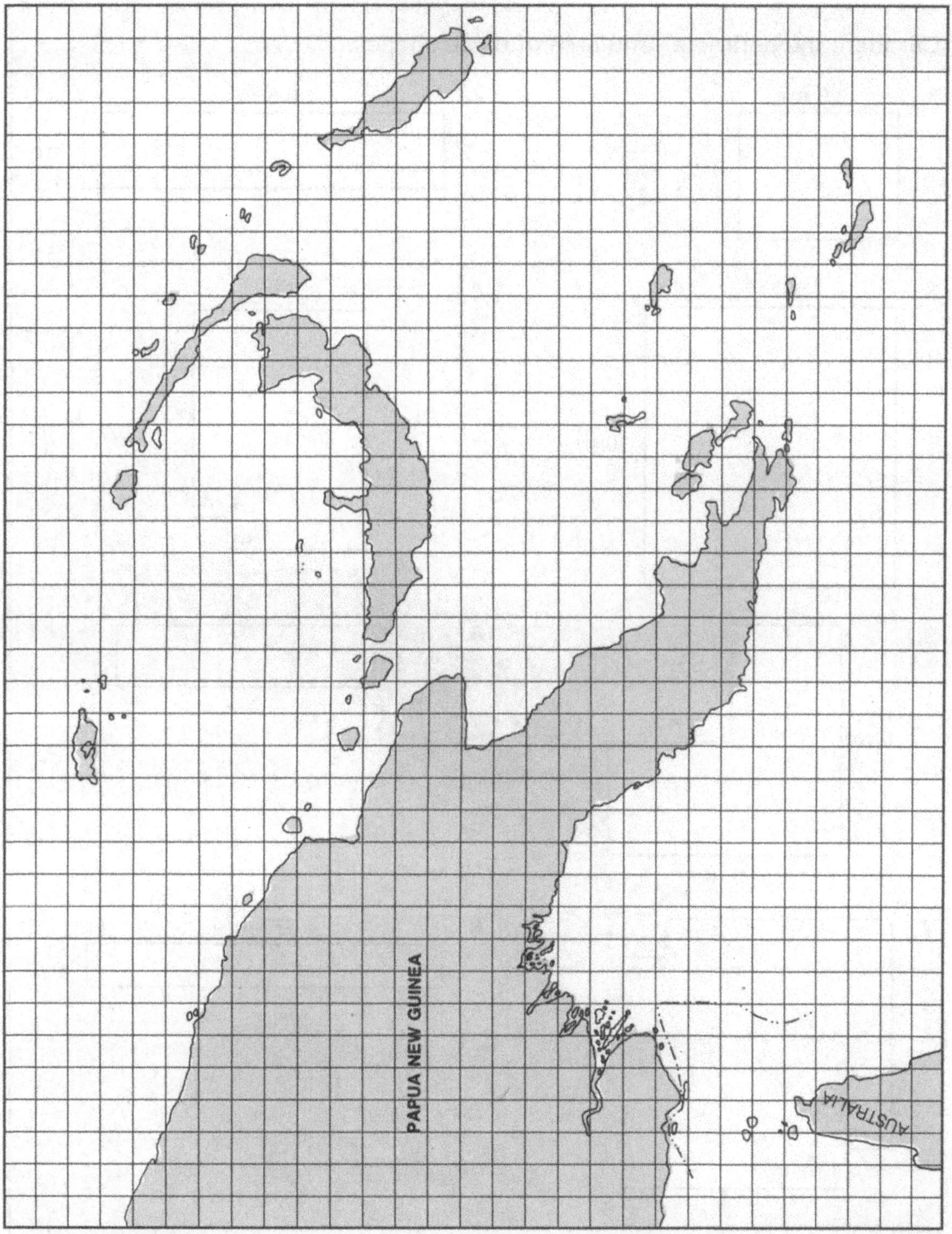

Calculating perimeter and area

Calculate the perimeter and area of these shapes.

1.

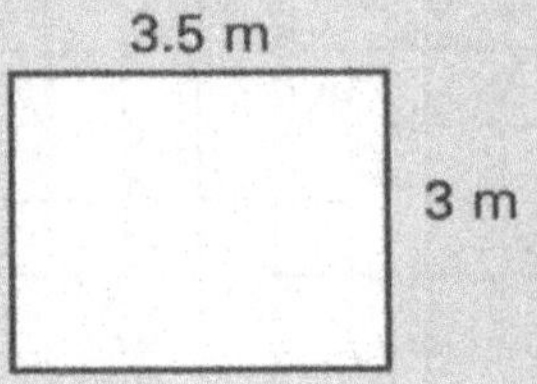

2.

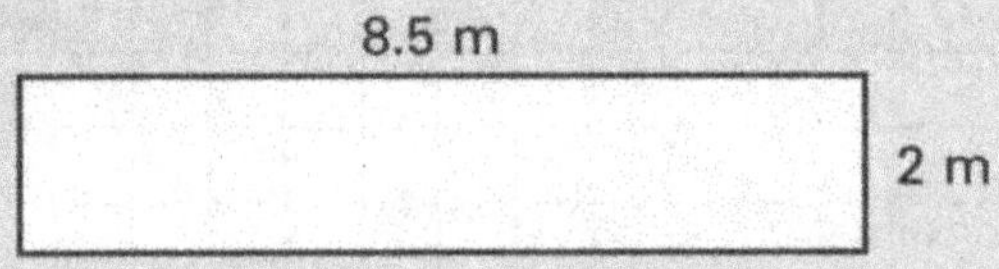

3.

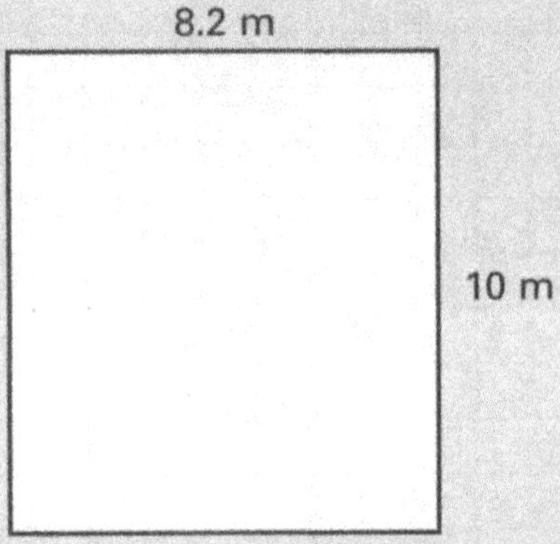

4.

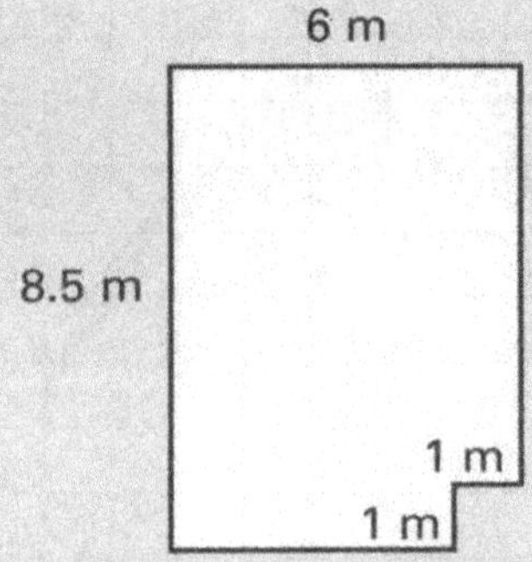

5.

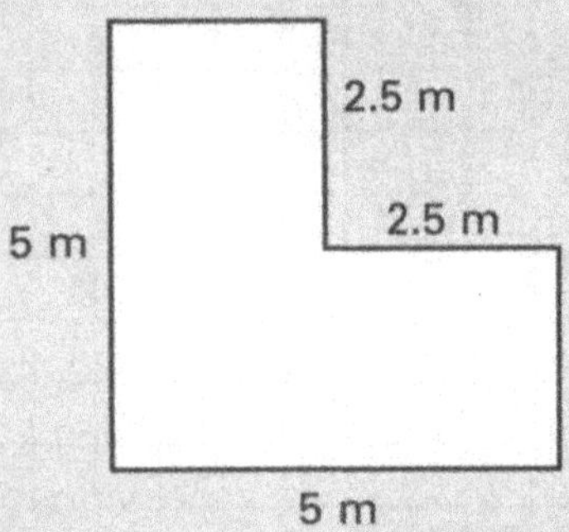

6.

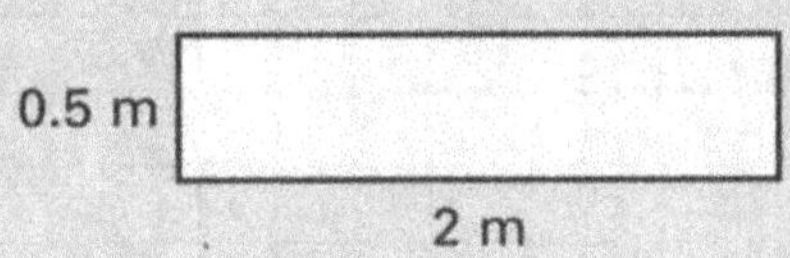

7.

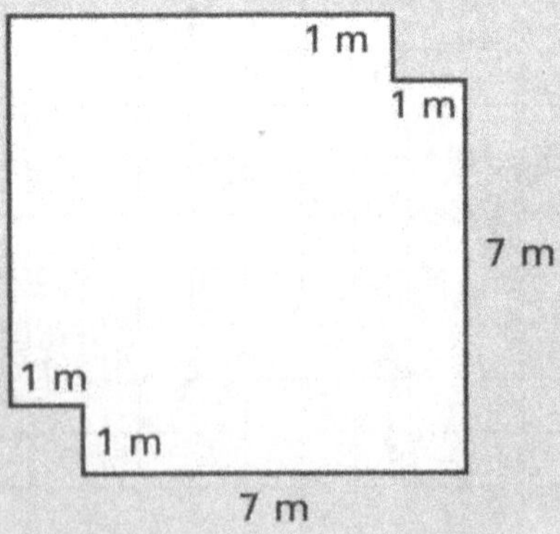

8.

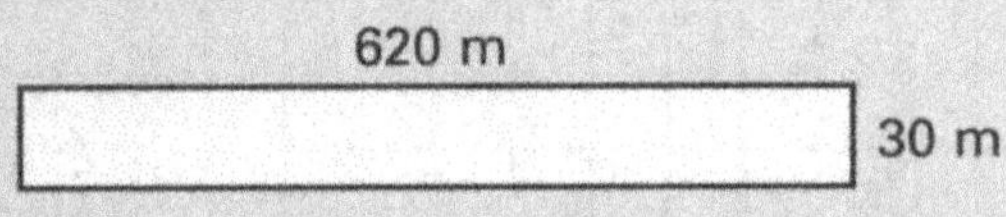

Length and area problems

1. What is the total length of 8 sticks each 5.2 m in length?
2. Land with a total area of 65 square metres is divided into four equal parts. What is the area of each part?
3. How much further is a throw of 31.50 m than 23.7 m?
4. What is the total height of the children in your row?
5. What is the floor area of your classroom?
6. If the perimeter of a rectangle is 30 cm, what could the area be?
7. If the area of a rectangle is 30 square centimetres, what could the perimeter be?
8. I am making a garden. I have enough fence material to make a fence 24 m long, including a gate. Which shape of garden will give me the most garden space?
9. Seri is keeping laying hens. Each hen needs 1.5 square metres ground space. How many hens can Seri fit into a chicken house 5 m long and 4 m wide?

10. **(a)** Walo is keeping meat birds. Each bird needs 0.5 square metres of ground space. Walo has a shed on the side of his house that is 1.5 m wide and 12 m long. How many meat birds can be put in the shed?

    (b) Walo only needs to fence three sides of the shed because one side is the wall of the house. What length of fencing wire does Walo need?

11. **✎ Challenge activity ✎**

I want to design a plantation that has an area of 1 hectare. What shapes can I use to make an area of 1 hectare? Which shape will need the shortest fence? If all sides must be straight lines at right angles to each other, what shape will be the shortest fence?

Looking at prisms and cylinders

1. Copy each solid shape into your book and write its name. Choose from the names given here.

cylinder   square prism   triangular prism   hexagonal prism   rectangular prism

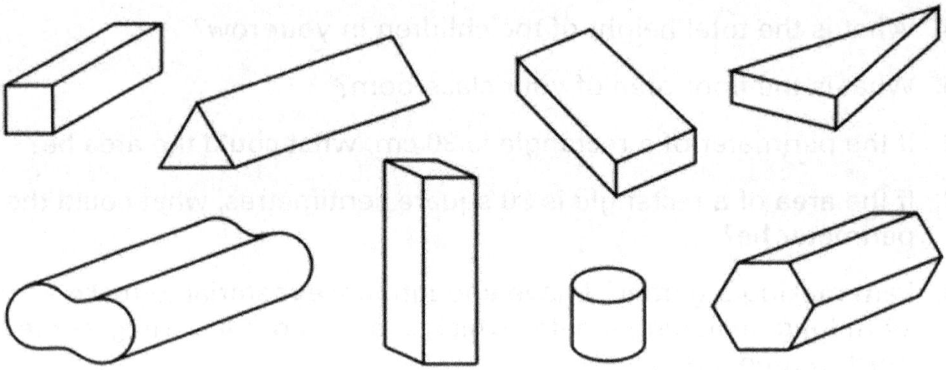

2. (a) Which of the above 5 shapes has the most edges?
   (b) Which has the fewest faces?
   (c) Which has the most vertices (corners)?

3. What shape could each set of faces make?

(a) 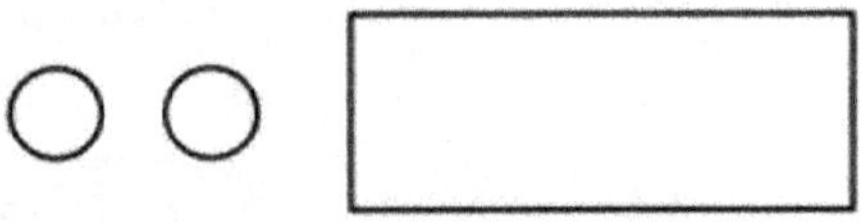

(b) 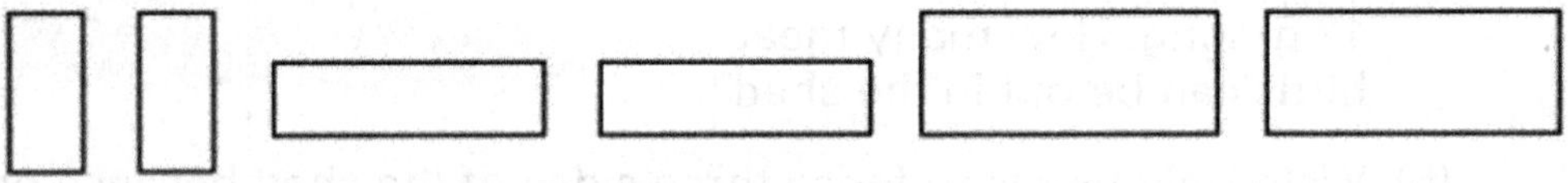

(c) 

(d) 

(e) 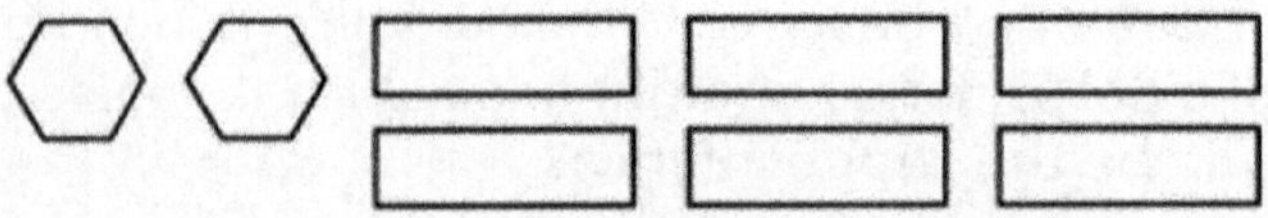

Where do we see pyramids?

Looking at 3-D shapes

1. Draw each solid shape in your book and write its name. Choose from the names given here.

triangular prism	rectangular prism	cylinder	cone
triangular pyramid	sphere	cube	square pyramid

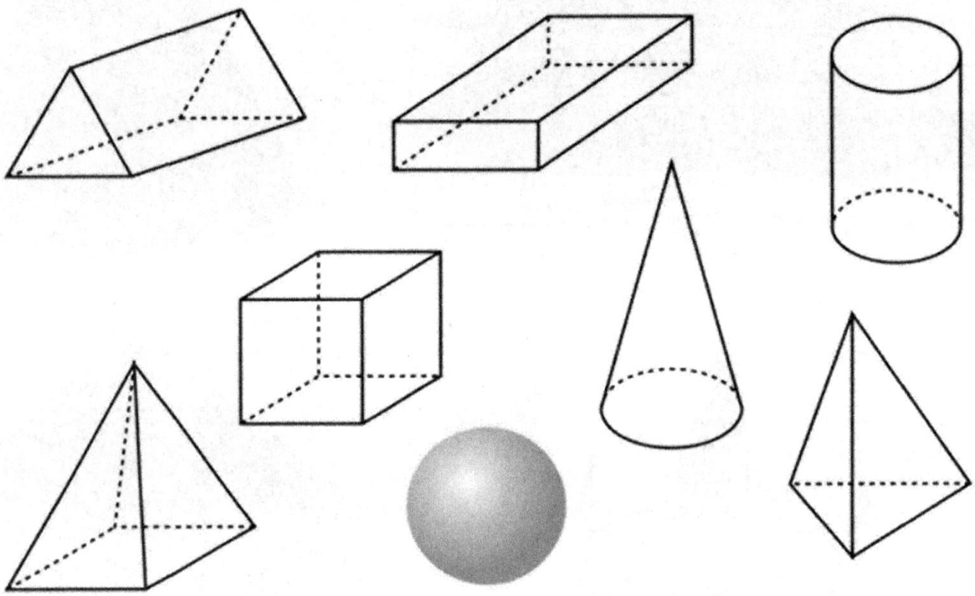

2. Draw something that is the same shape as each solid shown above.

3. What am I ?

   (a) I have one curved surface.
   I have one apex.

   (b) I have 4 faces and 4 vertices.

   (c) I have 6 faces. Each face is a square.

   (d) I have two circular ends.
   I have one curved surface.

   (e) I have 8 edges and 8 vertices.
   My faces are rectangular.

   (f) I have 5 faces. 4 of them are triangular.

Making solids from nets

Copy these nets exactly onto paper, cut them out and fold along the dotted lines. Paste the tabs to join each shape together.

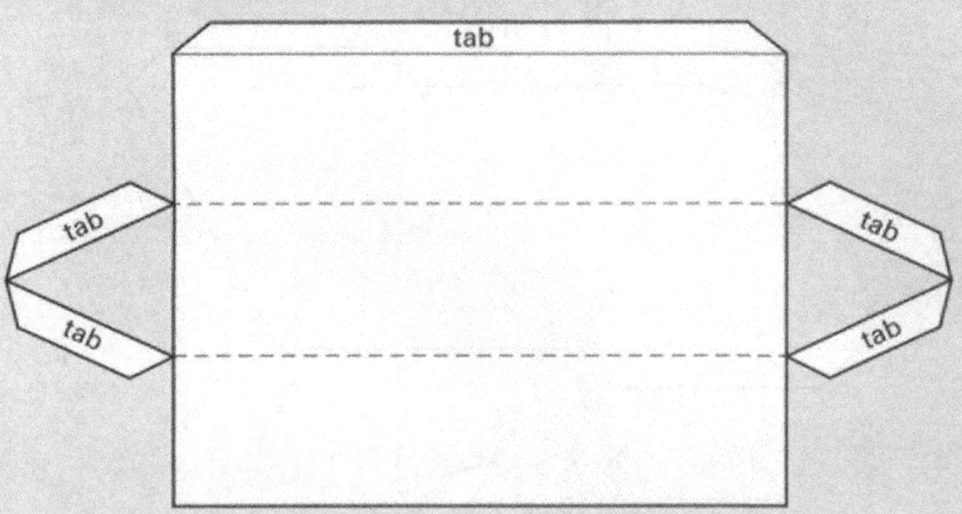

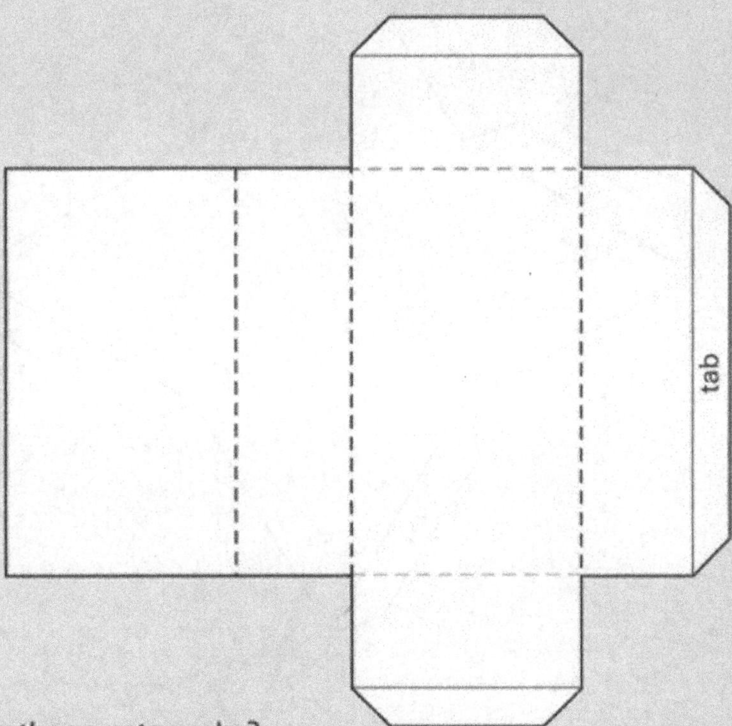

What shapes do these nets make?

Making solids from nets

Copy these nets exactly on to paper, cut them out and fold along the dotted lines. Paste the tabs to join each shape together.

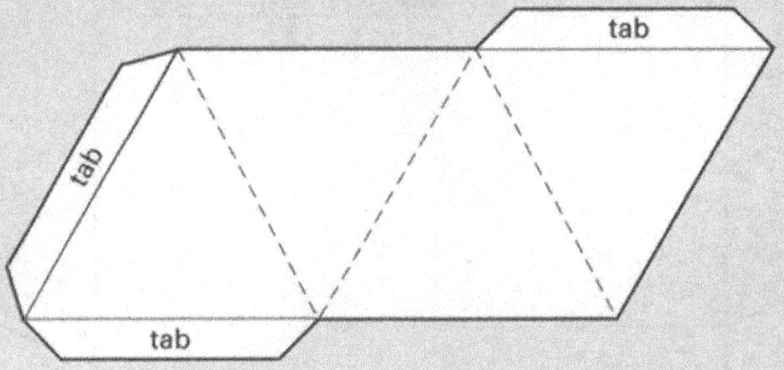

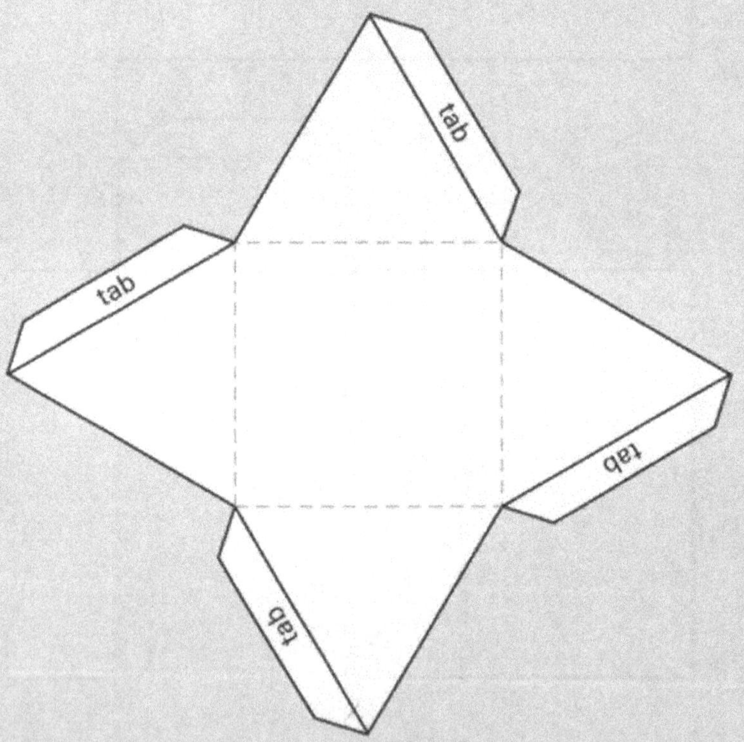

What shapes do these nets make?

Which net will make the shape?

Copy each shape into your book and draw the net/s that will make the shape. Make any that you are unsure of.

1.

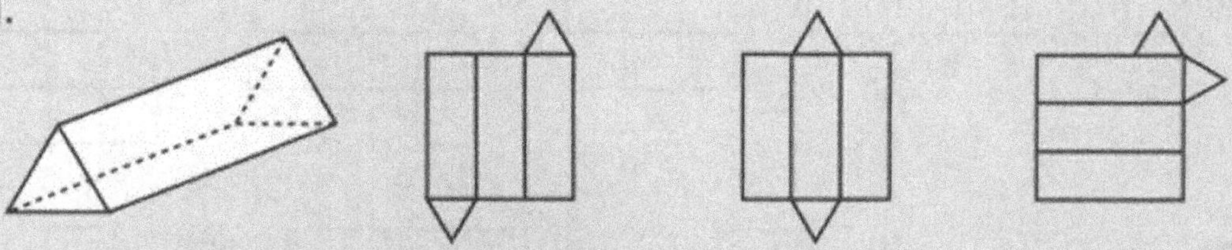

2.

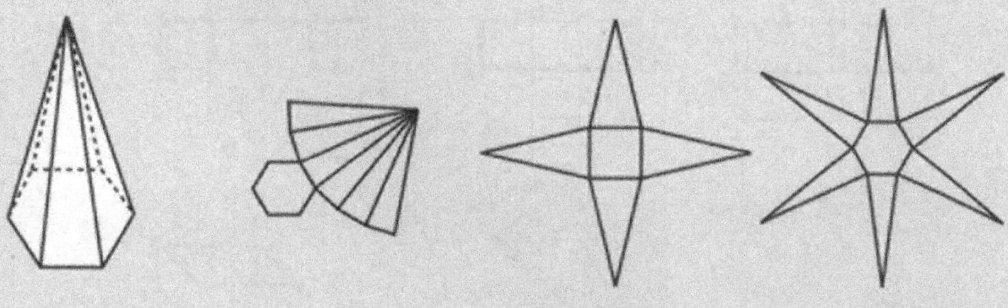

3.

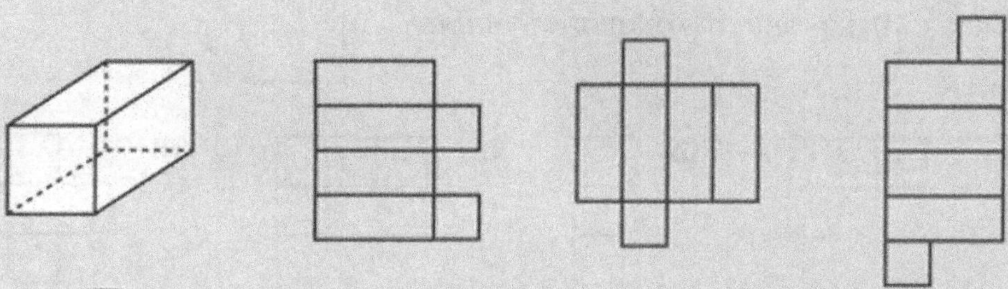

4.

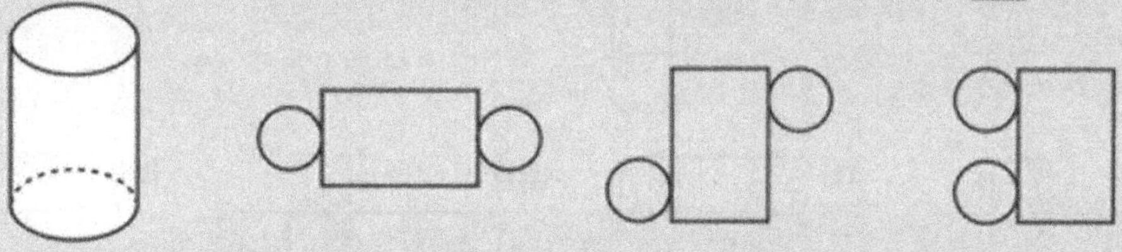

5.

How do you make a cube?

1. Which of these nets fold up to make a cube? If you are not sure, make them to find out.

(a)

(b)

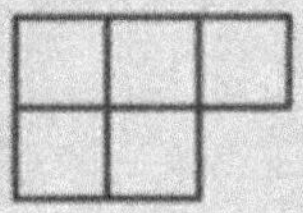

(c)

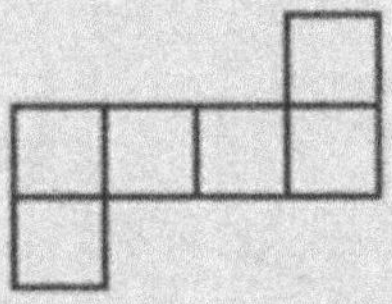

(d)

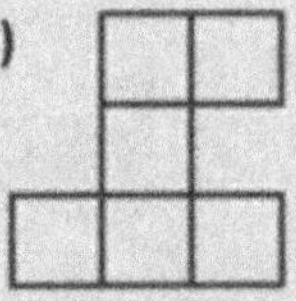

(e)

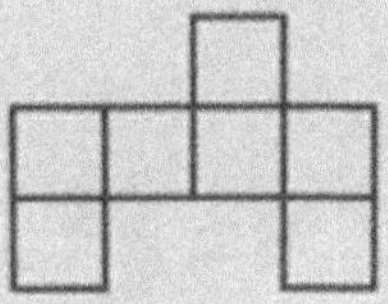

(f)

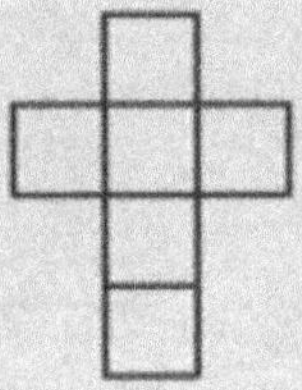

(g)

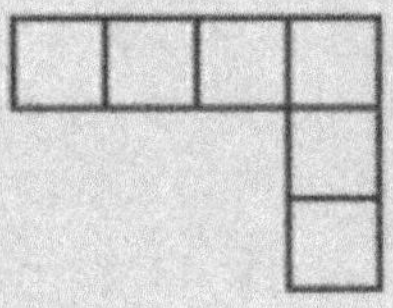

(h)

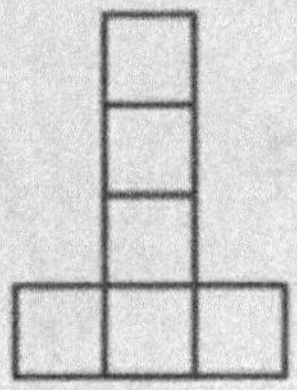

2. Which of these nets make an open cube?

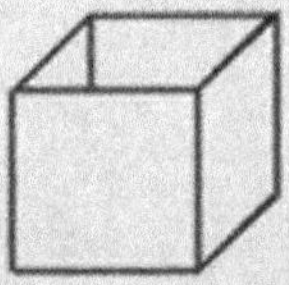

(a)

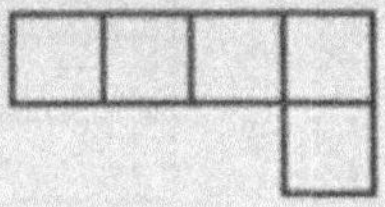

(b)

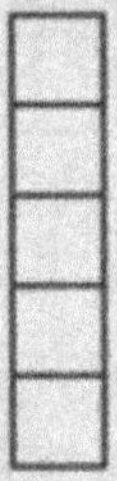

(c)

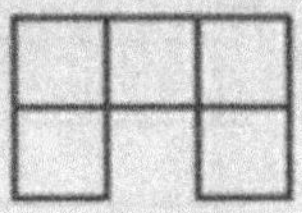

(d)

(e)

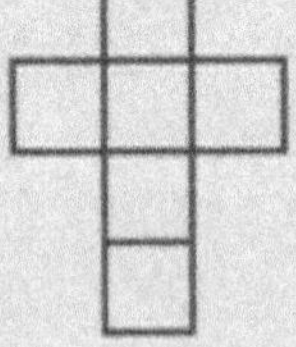

(f)

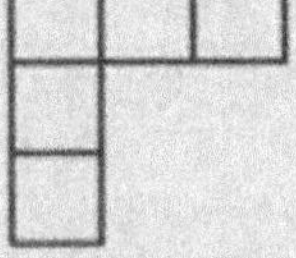

(g)

(h)

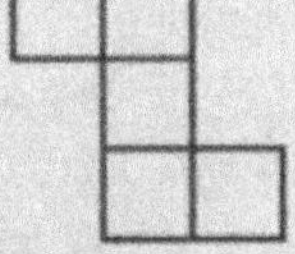

3. Draw your own net for a cube. Make sure you include tabs to join it together. When it is still flat, mark the dots to make a dice. Where do you need to place each set of dots so that when the dice is completed its opposite sides total 7?

Recording outcomes

What are the possible outcomes of the following experiments? Write them in your exercise book.

1. 

If you pick two cards at a time from these three cards, what are the possible combinations?

2. 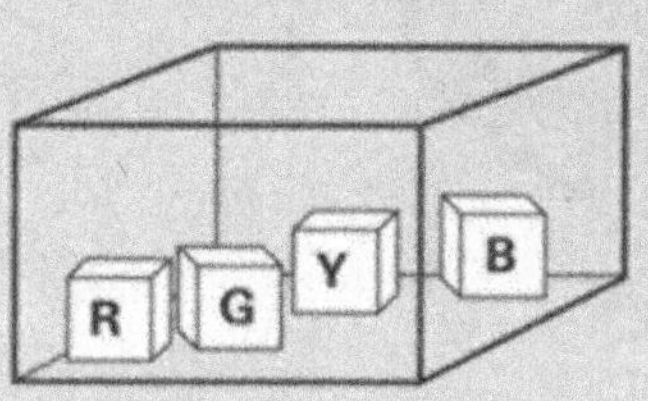

If you pick two cubes at a time what are the possible combinations?

3. 

If you choose two names out of the bag, what are the possibilities?

4. 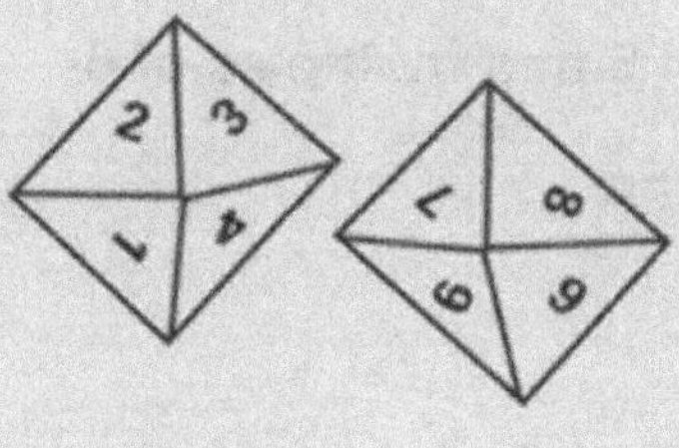

If you toss both dice together what are the possibilities?

5. 

If you toss the coin and one dice what are the possible outcomes?

6. 

If you toss three coins what are the possibilities?

Elements of sets

The elements of the set of primary colours can be written as: red, blue, yellow
or using a Venn diagram as shown:

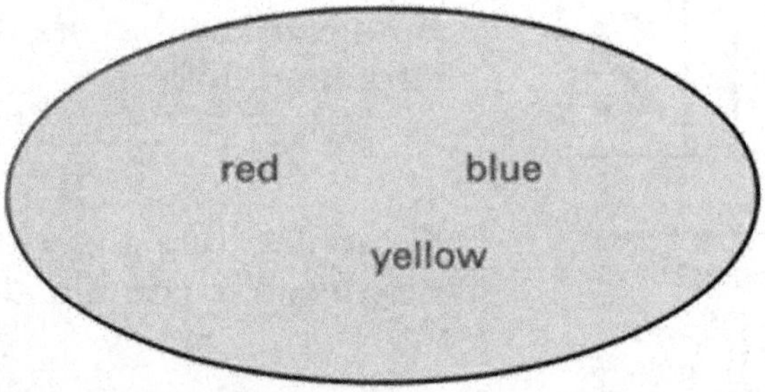

1. Write down the elements of the:

   (a) set of fruit that you ate last week
   (b) set of multiples of 3 below 20
   (c) set of factors of 30
   (d) set of even numbers between 60 and 80
   (e) set of names of shapes that have 4 sides
   (f) set of students in your class with the letter 't' in their name
   (g) set of all even numbers
   (h) set of all numbers greater than 100
   (i) set of neighbouring countries of PNG
   (j) set of your favourite numbers
   (k) set of numbers that are both multiples of 5 and multiples of 6

2. Use Venn diagrams to show the elements of the:

   (a) set of the first 5 square numbers
   (b) set of the letters of the alphabet
   (c) set of vegetables that you like to eat
   (d) set of months of the year
   (e) set of factors of 40
   (f) set of multiples of 5 that are less than 50
   (g) set of provinces in the Island Region of Papua New Guinea
   (h) set of fruit growing near your house
   (i) set of people that live in your home
   (j) set of prime numbers less than 50
   (k) set of numbers that are factors of 20 and also factors of 24.

Describing sets

Write a sentence to describe these sets using words.

1. 0, 1, 2, 3, 4, 5, 6, 7, 8, 9
2. Western, Gulf, NCD, Central, Oro, Milne Bay, Southern Highlands
3. 1t, 2t, 5t, 10t, 20t, 50t, K1
4. 4, 8, 12, 16, 20, 24, 28
5. Monday, Tuesday, Wednesday, Thursday, Friday
6. 1, 2, 3, 4, ...
7. Madang, Wewak, Vanimo, Lae
8. 10, 20, 30, 40, ...

9.

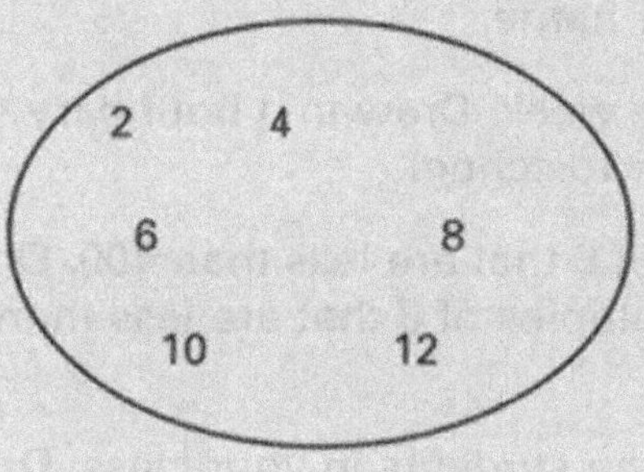

10.

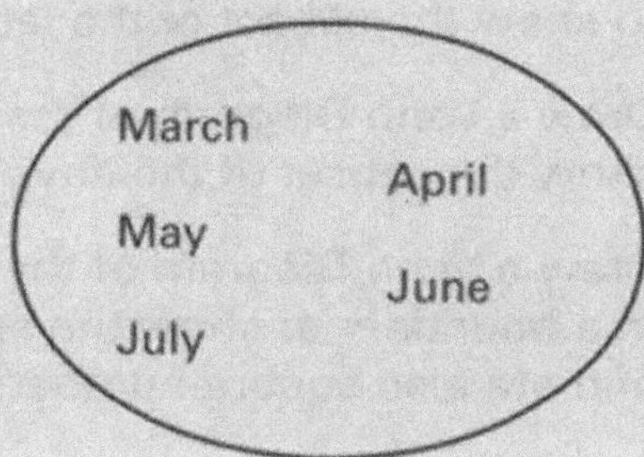

11.

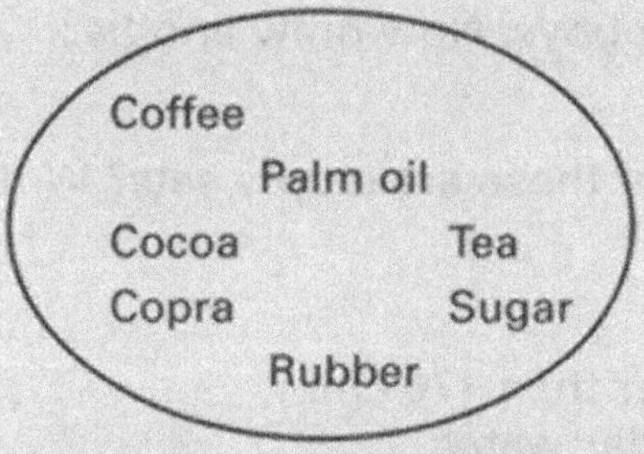

12.

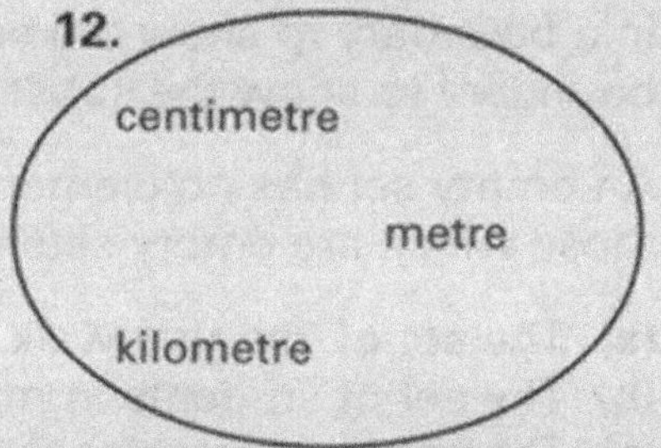

13.

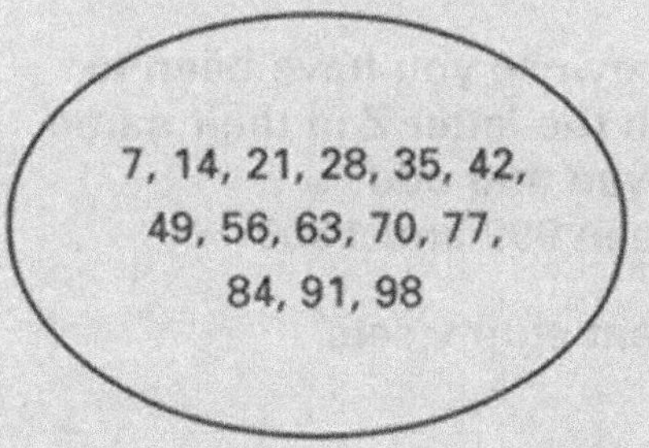

14.

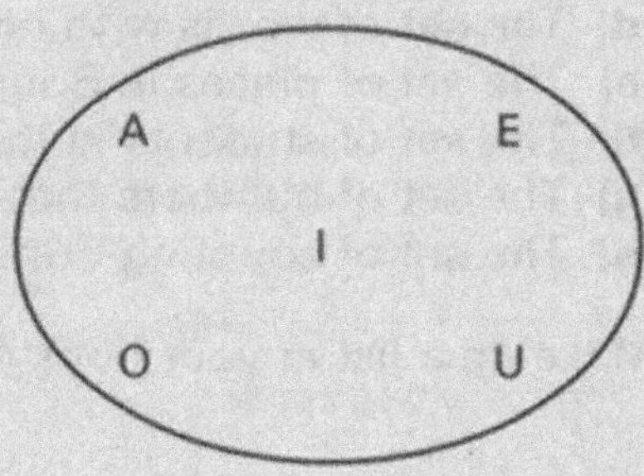

Sets and subsets

This is the set of the letters in the words Papua New Guinea.

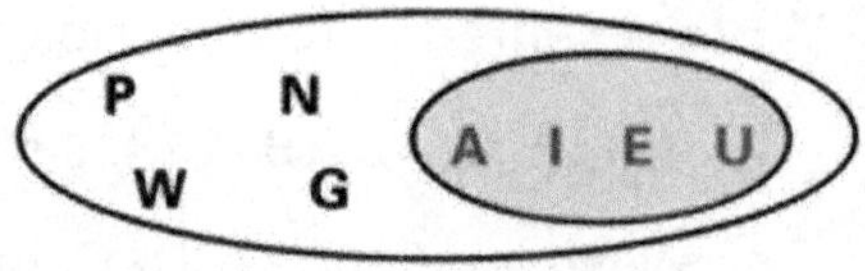

Letters in the words Papua New Guinea

The **subset** of the vowels in the words Papua New Guinea is shown.

1. Draw a Venn Diagram of the months of the year. Draw in a boundary to show the subset of the months that you do not go to school.

2. Draw a Venn Diagram of the factors of 36. Draw in a boundary to show the subset of the factors of 12.

3. Draw a Venn Diagram of the first 10 square numbers. Draw in a boundary to show the subset of the first 5 odd square numbers.

4. Draw a Venn Diagram of the letters of the alphabet. Draw in a boundary to show the subset of the letters in your name.

5. Draw a Venn Diagram of the days of the week. Draw in a boundary to show the subset of the days that you go to school.

6. Draw a Venn Diagram of the multiples of 8 that are less than 100. Draw in a boundary to show the subset of multiples of 8 that are less than 100 and are also square numbers.

7. Draw a Venn Diagram of the names of the students in your class. Draw in a boundary to show the subset of the boys. Now draw another boundary to show the subset of girls.

8. An empty set has no elements. Which of these are empty sets? Write those which are empty sets in your book.

   (a) The set of fish that walk.
   (b) The set of students in my class taller than 170 cm.
   (c) The set of vegetables that grow under water.
   (d) The set of words with no vowels.
   (e) The set of places in Bougainville Province you have been to.
   (f) The set of students in the class with the letter Z in their name.
   (g) The set of numbers that are both even and odd.
   (h) The set of counting numbers between 999 and 1 000.

9. Make up a list in your book of 10 different empty sets.

Intersecting sets

I want to show two sets. The first set is the even numbers less than 10. The second set is the multiples of 3 less than 10. The Venn diagram of these sets might look like:

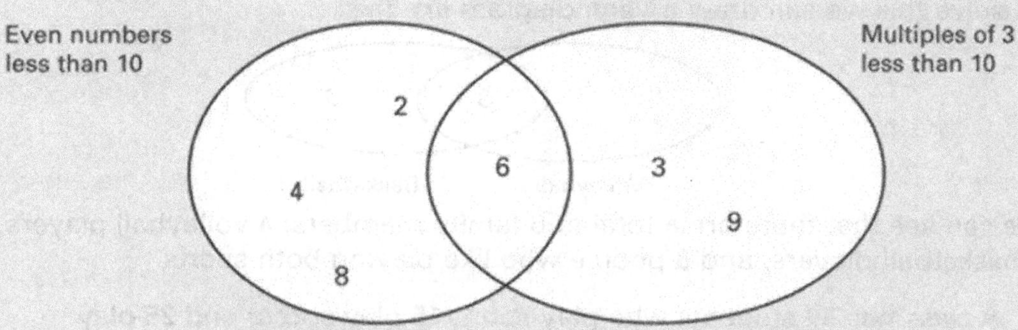

1. Write down the elements that are members of both the set of multiples of 4 that are less than 100 and the set of multiples of 5 that are less than 100.
2. Write down the elements that are members of both the set of vegetables grown near your place and the set of food that you like.
3. Write down the elements that are members of both the set of factors of 24 and the set of factors of 36.
4. Write down the elements that are members of both the set of factors of 16 and the set of factors of 20.
5. Draw a Venn diagram to show both the set of factors of 100 and the set of square numbers less than 100.
6. Draw a Venn diagram to show both the set of students in your class taller than you and the set of girls, and show which elements are in both sets.
7. Draw a Venn diagram to show both the set of students in your class with the letter 'a' in their name and the set of girls, and show which elements are in both sets.
8. Draw a Venn diagram to show both the set of factors of 36 and the set of factors of 60, showing which elements are in both sets.

Story problems for sets

Example: There are 6 people in my family. 4 people like volleyball, and 5 people like basketball. How many people like both volleyball and basketball?

To solve this we can draw a Venn diagram like this:

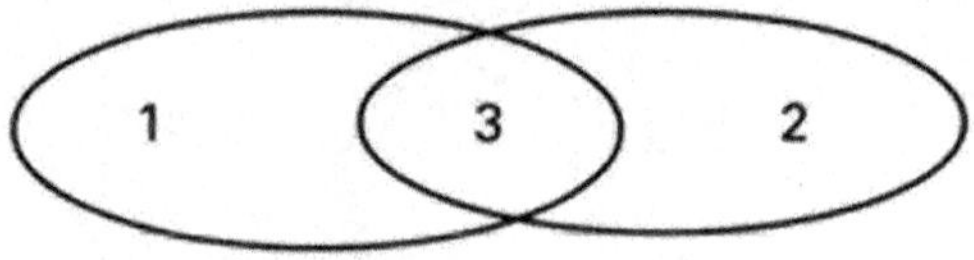

We can see that there are a total of 6 family members; 4 volleyball players, 5 basketball players, and 3 people who like playing both sports.

1. A class has 30 students who play sport. 15 play soccer and 25 play softball. How many play both soccer and softball?
2. There are 8 months with the letter r, and 6 months with the letter a. How many months are there with both letters?
3. A class has 25 students who like fruit. 20 of them like tomatoes and 12 like oranges. How many like both?
4. A class has 20 students who like to read. 10 students like to read stories and 15 students like to read the Bible. How many like to read stories and the Bible?
5. A class has 25 students. In addition to their own language, 18 of them speak Tok Pisin and 12 speak Motu. How many speak both Tok Pisin and Motu?
6. A school has 225 students who like to cook. 150 of them like to cook kaukau and 120 like to cook rice. How many like to cook both?

7. Challenge activity

In a room there are 12 boys and 25 students. How many people might there be in the room? List all the possible answers.

8. Challenge activity

In a room there are 12 basketball players and 25 people who play volleyball. How many people might there be in the room? List all the possible answers.

Finding symmetry

1. Do these polygons have horizontal symmetry, vertical symmetry, or both? Use a mirror to find out.

(a)

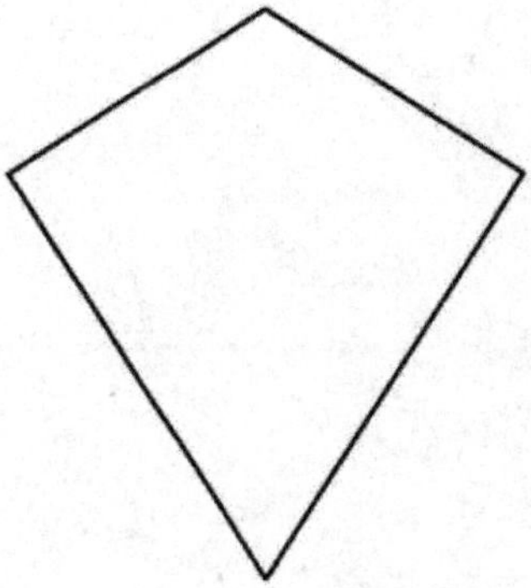

(b)

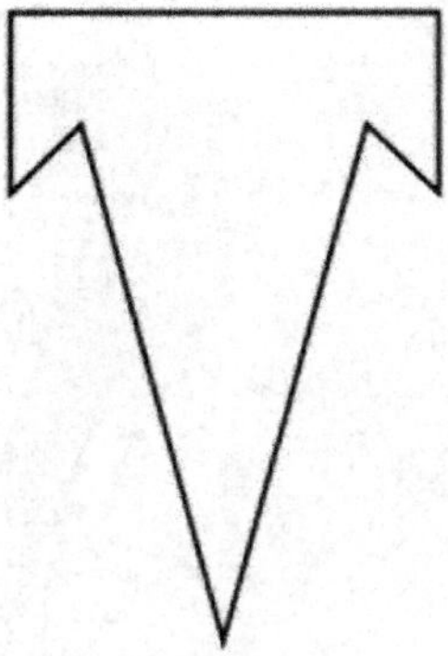

(c)

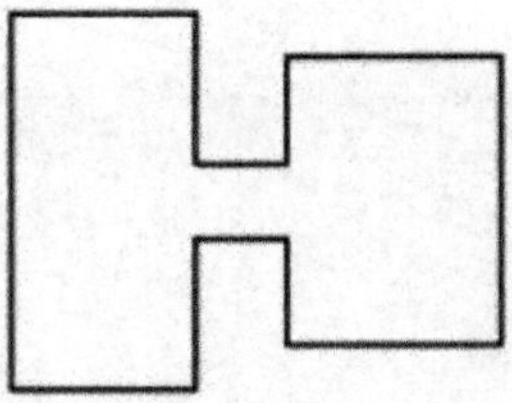

(d)

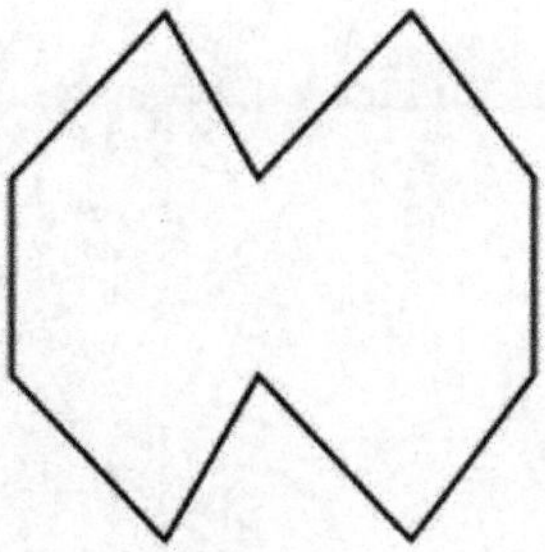

(e)

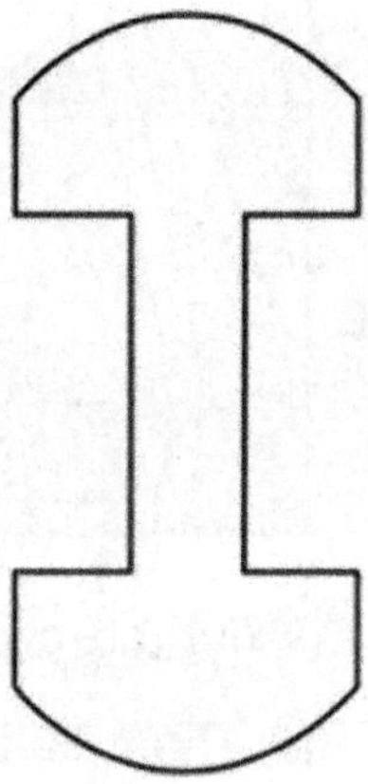

(f)

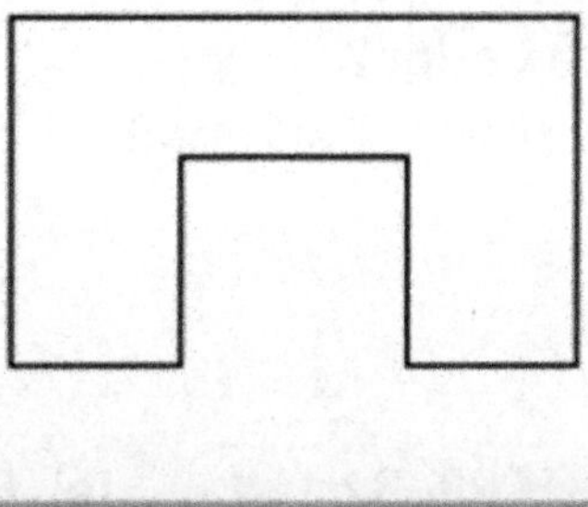

(g)

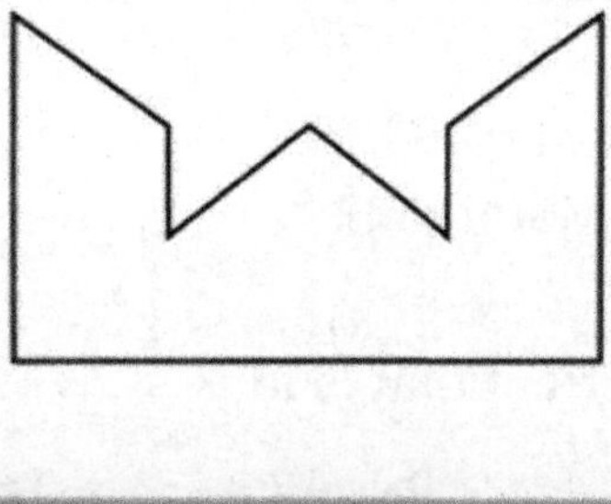

2. **Challenge activities**

(a) Draw a polygon that has only horizontal symmetry.
(b) Draw a polygon that has only vertical symmetry.
(c) Draw a polygon that has both horizontal and vertical symmetry.

Using a map of Port Moresby

PORT MORESBY CITY CENTRE

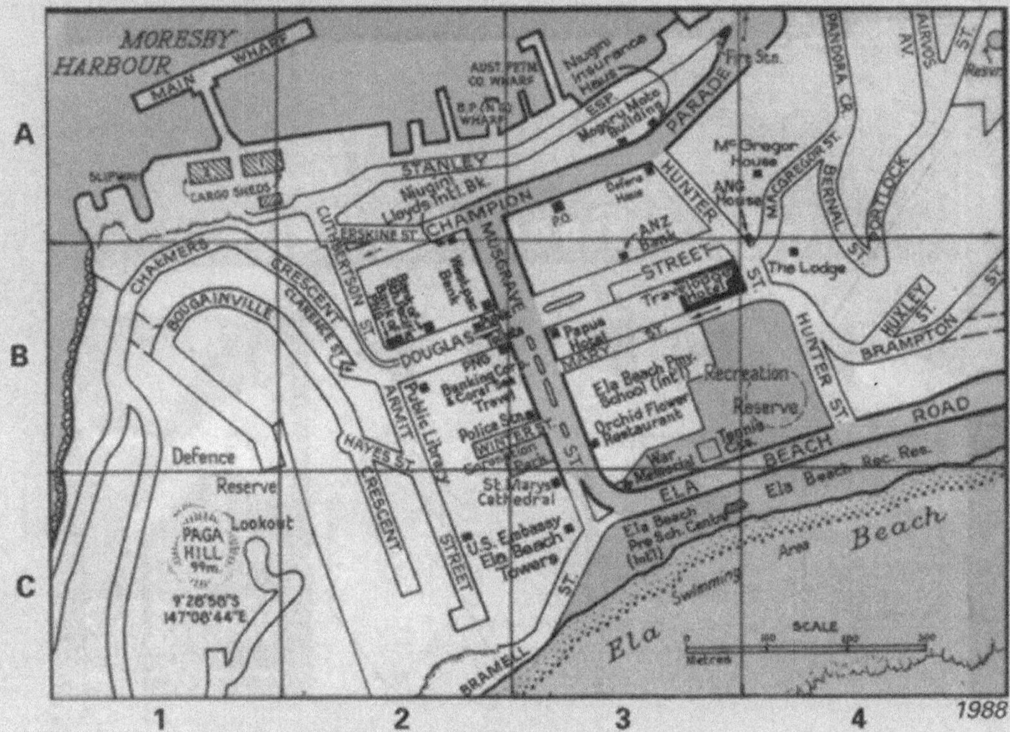

1. Which co-ordinates would be used for:

   (a) Paga Hill?
   (b) Defence Haus?
   (c) Ela Beach Pre-School Centre?
   (d) Post Office?
   (e) Airvos Av.?
   (f) The Lodge?
   (g) Bank of PNG?
   (h) St. Mary's Cathedral?
   (i) War Memorial?

2. Write a street that is in:

   (a) B4 (b) A2 (c) A4 (d) B2 (e) C3

3. Which squares is Ela Beach in?

4. Name as many features in B3 as you can.

A map of Lae

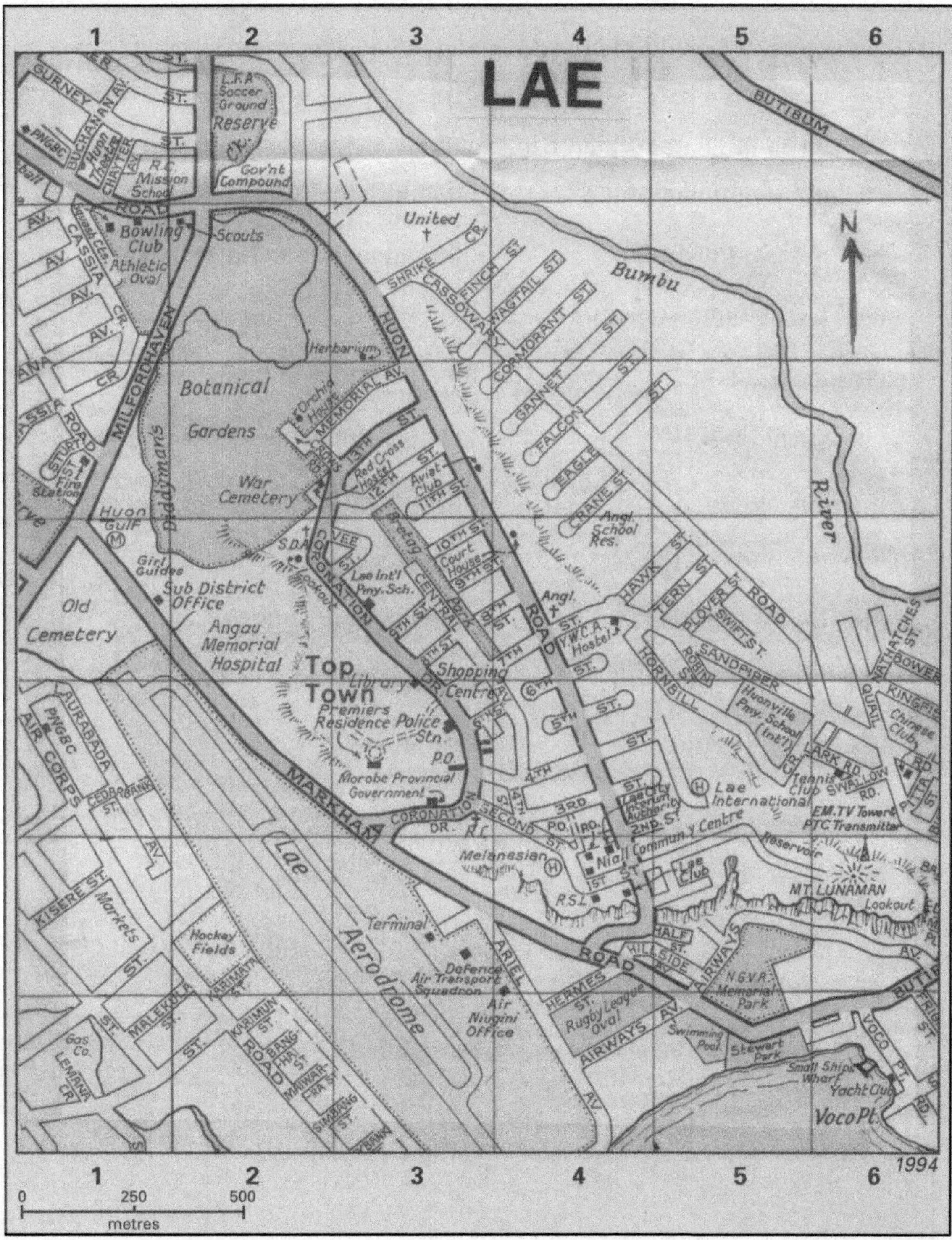

LAE
1
2
3
4
5
6
A
B
C
D
E
F
G
GURNEY
BUCHANAN AV.
Huon Theatre
CHAYTER
R.C. Mission School
L.F.A. Soccer Ground
Reserve
Gov'nt Compound
ROAD
Bowling Club
Scouts
Athletic Oval
CASSIA
MILFORDHAVEN
United
SHRIKE CR.
CASSOWARY
FINCH ST.
WAGTAIL ST.
CORMORANT ST.
GANNET ST.
FALCON ST.
EAGLE
CRANE ST.
Bumbu
River
BUTIBUM
N
HUON
Herbarium
Botanical Gardens
Orchid House
MEMORIAL AV.
13TH ST.
Red Cross Hostel
12TH ST.
Aviat Club
11TH ST.
War Cemetery
Didiyman's
Huon Gulf
Fire Station
STURT
S.D.A.
CORONATION
Girl Guides
Sub District Office
Old Cemetery
Angau Memorial Hospital
Lookout
Lae Int'l Pmy. Sch.
10TH ST.
Court House
9TH ST.
8TH ST.
7TH
Angl.
Angl. School Res.
HAWK ST.
TERN ST.
PLOVER ST.
SWIFT ST.
ROAD
Y.W.C.A. Hostel
HORNBILL
ROBIN ST.
SANDPIPER
Huonville Pmy. School (Int'l)
Top Town
Library
Shopping Centre
Premiers Residence
Police Stn.
P.O.
Morobe Provincial Government
6TH
5TH ST.
4TH
3RD
2ND ST.
1ST ST.
Lae City Interim Authority
Lae International
Niall Community Centre
Melanesian
R.S.L.
Lae Club
Reservoir
MT. LUNAMAN
Lookout
EM.TV Tower
PTC Transmitter
Tennis Club
LARK RD.
SWALLOW RD.
QUAIL
NATHATCHES ST.
BOWER
KINGFIS
Chinese Club
MARKHAM
CORONATION DR.
SECOND ST.
R.C.
LAURABADA
AIR CORPS
PNGBC
CEDAR
BANK
KISERE ST.
Markets
Hockey Fields
Lae
Aerodrome
Terminal
Defence Air Transport Squadron
Air Niugini Office
ARIEL
HERMES ST.
Rugby League Oval
AIRWAYS AV.
HILLSIDE AV.
HALF ST.
AIRWAYS
N.G.V.R. Memorial Park
Swimming Pool
Stewart Park
Small Ships Wharf
Yacht Club
Voco Pt.
VOCO PT. RD.
MALEKULA ST.
Gas Co.
LEMANA CR.
KARIMUN ST.
BANGHAI ST.
MAIWARA ST.
SIMBANG ST.
1994
0
250
500
metres

Using a map of Lae

Use the map of Lae on the previous page to answer these questions.

1. Which co-ordinates would be used for:

(a) the swimming pool?

(b) the Yacht Club?

(c) the Old Cemetery?

(d) Angau Memorial Hospital?

(e) Mt. Lunaman Lookout?

(f) Huonville Primary School?

(g) the R.S.L.?

(h) the Chinese Club?

(i) Orchid House?

(j) the Premier's Residence?

(k) 5th Street?

(l) Stewart Park?

(m) the Herbarium?

(n) Lae Club?

(o) Shrike Crescent?

(p) the Court House?

(q) the Police Station?

(r) Lae International Primary School?

(s) Swift Street?

(t) the Library ?

2. Name something in each of these squares:

(a) A2

(b) B1

(c) B5

(d) G1

(e) F2

(f) C3

3. What co-ordinates could be used for the Bumbu River?

4. What co-ordinates could be used for the Lae airport?

Understanding scale

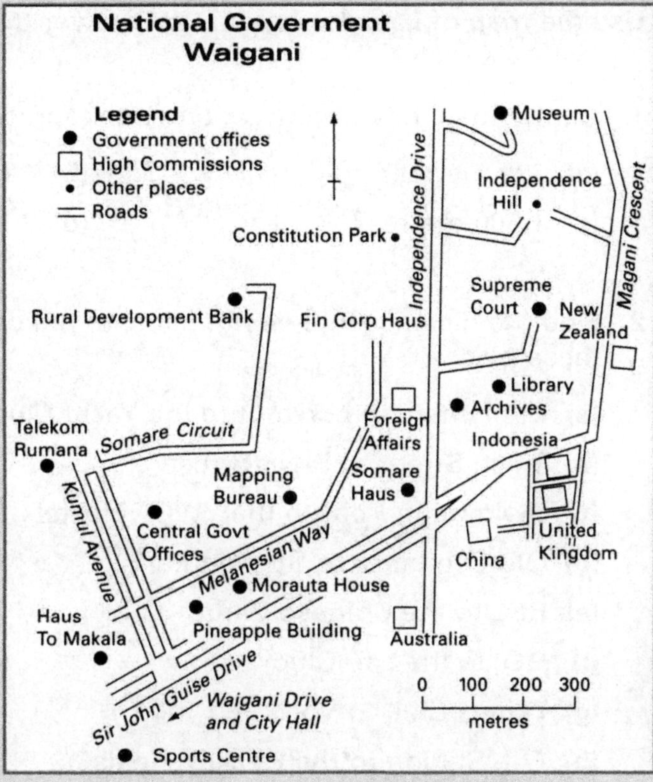

1. On this map the scale shows that 2 cm = ☐ metres
2. What does 1 cm on the map represent?
3. How many metres on the map does 6 cm represent?
4. What does 10 cm represent?
5. How many metres does $\frac{1}{2}$ cm represent?
6. Use the scale to calculate the distance in a straight line between these places:
   - **(a)** Parliament House to the Supreme Court
   - **(b)** Pineapple Building to the Museum
   - **(c)** Archives to the Library
   - **(d)** Sports Centre to the Mapping Bureau
   - **(e)** Bank to the Supreme Court
   - **(f)** Morauta House to Somare Haus
7. How many metres would you walk if you went from one end of Independence Drive to the other end?
8. If you walked from Telekom Rumana down Kumul Ave. and turned into Melanesian Way and continued until you reached the Australian High Commission, how far would you have walked?

Calculating distances and costs in Lae

Use the map of Lae on page 171 to answer these questions.

1. On the map, how many centimetres represent:

   (a) 500 metres?
   (b) 1 kilometre?
   (c) 1 500 metres?
   (d) 2 kilometres?

2. Find the distance in a straight line, to the nearest 100 metres, between these places.

   (a) Mt. Lunaman Lookout to the Yacht Club
   (b) Finch Street to Swift Street
   (c) Swimming Pool to the YWCA Hostel
   (d) Old Cemetery to Stewart Park
   (e) RSL to the Chinese Club
   (f) P.O. to the Lae Club
   (g) Tennis Club to Voco Pt.
   (h) Fire Station to the Hockey Fields

3. Use string and the scale to find the distance from:

   (a) one end of Huon Road to the other end.
   (b) the northern end of Milfordhaven St. to the Rugby League oval on Markham Road.

4. If a taxi cost K5 per kilometre, how much would you pay to travel from:

   (a) Huonville Primary School along Sandpiper Road and 7th Street to the Shopping Centre on Coronation Drive?
   (b) N.G.V.R. Memorial Park on G5 along Markham Road, turning right at Coronation Drive until you reach 12th Street?

5. A taxi fare cost K15. Where could you have travelled?

How many laps?

1. A running track is 400 metres per lap.

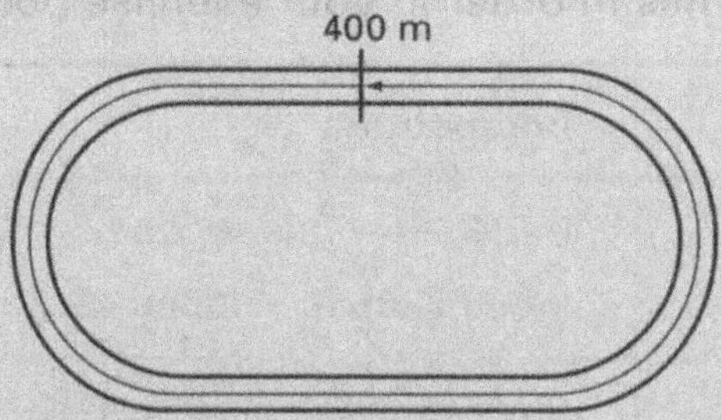

How many laps do competitors run in:

(a) 200 metres (b) 400 metres (c) 800 metres
(d) 1 500 metres (e) 3 000 metres (f) 5 000 metres
(g) 10 000 metres (h) 20 kilometres?

2. The pool is 50 metres long.

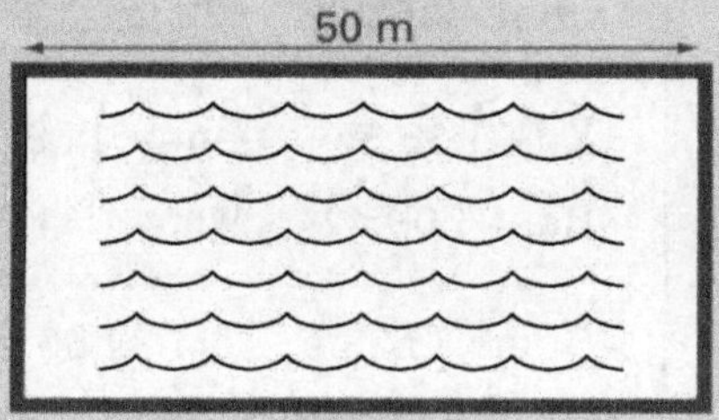

How many lengths do competitors swim in:

(a) 100 metres (b) 200 metres
(c) 400 metres (d) 1 500 metres?

3. The pool is 25 metres long.

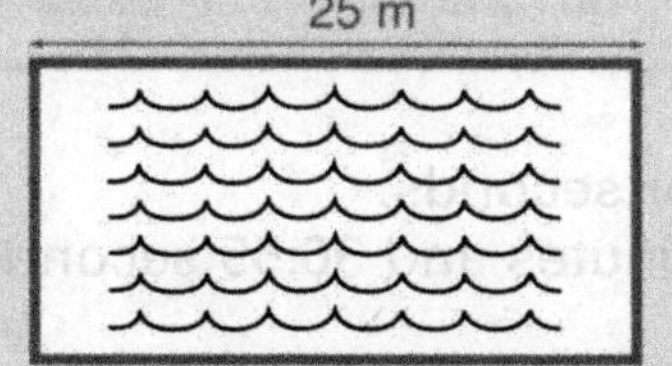

How many lengths do competitors swim in:

(a) 100 metres (b) 200 metres
(c) 400 metres (d) 1 500 metres?

Who came first?

Look at the results given for each event and work out who came first. Record the other names in order in your exercise book.

Event	Competitors
Discus	Kari 45 m Ella 47.25 m Ari 46.88 m Peter 47.65 m Julius 45.2 m Tom 46.5 m
100 metres sprint	Harry 15.6 s Lina 14.88 s Martin 14.94 s Gana 15.63 s Bai 14.49 s Elsie 14.05 s
Long jump	Joe 4.87 m Riki 5.05 m Jean 4.67 m Wari 4.92 m Kaia 5.23 m Nauna 5.16 m
800 metres	Sam 2:36.55 Tau 2:47.18 Tami 2:42.36 Louis 2:53.08 Philip 2:39.39 Emoto 2:47.23
100 metres freestyle	Kua 1:16.86 Jenny 1:08.76 Wari 1:15.33 Reni 1:09.42 Mata 1:11.24 Gana 1:21.45
50 metres freestyle	Suari 33.23 s Lia 32.56 s Peter 36.08 s Vavine 32.38 s Vagi 33.58 s Piawin 35.19 s
Shot put	Tami 15.63 m Biro 17.32 m Harry 15.07 m Hitolo17.47 m Galamo 16.28 m Tabua 17.50 m
High jump	Peter 1.63 m Regina 1.59 m Tau 1.71 m Bai 1.65 m Jean 1.56 m Charles 1.68 m

Note: 's' means second or seconds.
2:36.55 means 2 minutes and 36.55 seconds.

Sports graphs 1

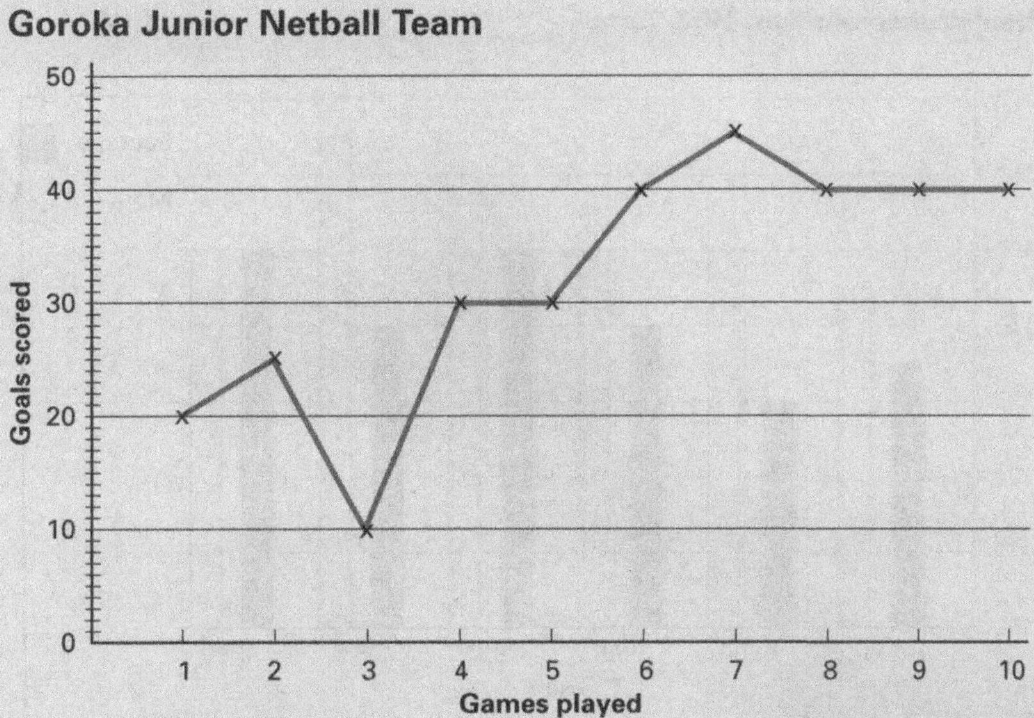

1. In what match did the team score the least number of goals?
2. In which matches were the team most consistent?
3. How many goals did the team score in the sixth match?
4. When did the team make their highest score?
5. Do you think the team improved towards the end of the season? What makes you think this?

Sports graphs 2

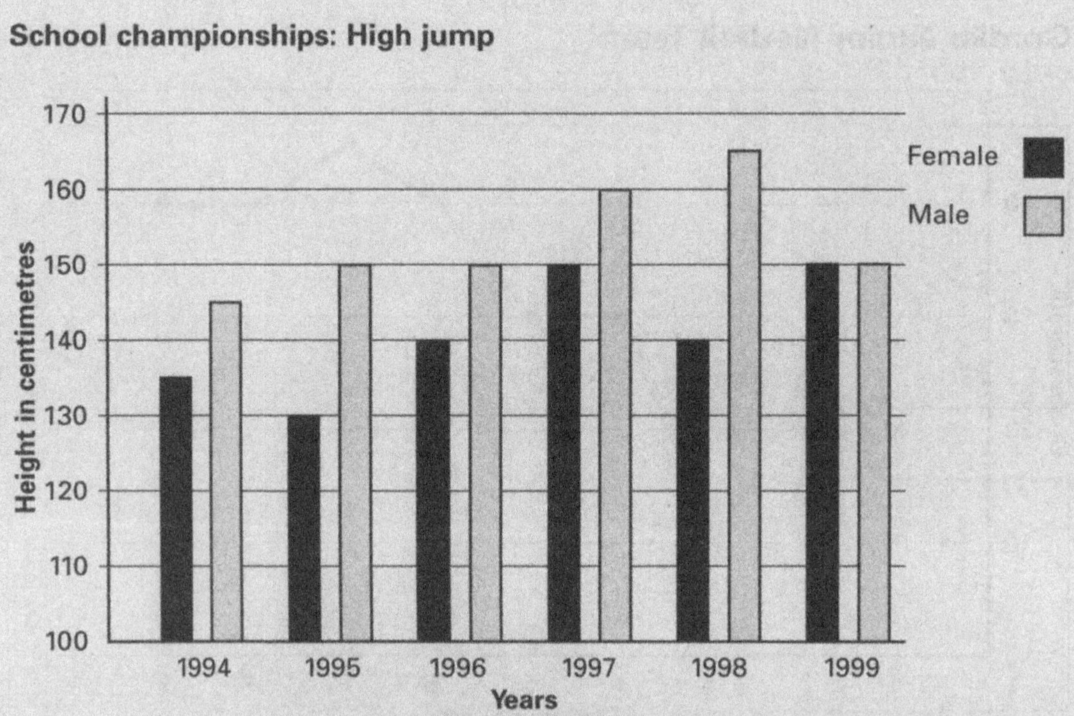

1. What height was the best jump by a female?
2. In which year did a male jump 1 m 65 cm?
3. In which year did the winning male and female jump the same height?
4. How high did the winning male jump in 1997?
5. Is there a year when the winning jump by a female was higher than for a male?
6. What was the worst winning performance by a male? a female?

The cost of sporting equipment

Calculate the cost of equipment for each of the sports and record them in order from most expensive to least expensive.

Netball		Soccer	
Net	K20	Shin pads	K9
Ball	K40	Ball	K45
Ring	K50	Boots	K75
Bib	K20	Socks	K15

Hockey		Softball	
Hockey stick	K25	Bat	K150
Ball	K20	Ball	K25
Shin pads	K9	Glove	K30

Rugby		Volleyball	
Ball	K50	Ball	K35
Boots	K75	Net	K90
Socks	K15	Whistle	K2

Sporting goods

Calculate the total cost of these orders for sporting equipment.

		Malibu Primary School		Yengep Primary School	
Item	Price per unit	Number required	Total	Number required	Total
Softball	K25	3		2	
Softball glove	K30	4		2	
Softball bat	K150	3		1	
Volleyball	K35	6		4	
Volleyball net	K90	4		2	
Basketball	K60	4		4	
Soccer ball	K45	2		4	
Rugby ball	K50	2		6	
Hockey ball	K20	2		6	
Hockey stick	K25	12		6	
Netball	K40	4		2	
Netball net	K20	2		1	
Netball ring	K50	14		1	
Netball bib	K20	4		0	
Whistle	K2	2		2	
Stopwatch	K120	2		1	
		TOTAL		TOTAL	

Which school spent the most money?

Sporting groups

Show the data in these tables in different ways.

1.

	Plays hockey	Doesn't play hockey	
Girls	Ella, Jenny, Lia, Galena, Kara	Alice, Naomi, Jean	(a) Does Sam play hockey? (b) How many girls play hockey? (c) How many boys were surveyed?
Boys	Charles, Kai, Joe	Suari, Philip, Sam, John	

2. Children's sport preferences

Soccer	𝍸 I	(a) How many more children play hockey than volleyball? (b) How many children were surveyed? (c) How many fewer children play soccer than basketball?
Basketball	𝍸 𝍸 III	
Hockey	𝍸 𝍸 𝍸 I	
Volleyball	𝍸 𝍸 I	

3. Triple jump results

Country	Distance in metres	
USA	13.54	(a) Which country's triple jump record is less than 13 metres? (b) What country does the athlete who jumped the longest distance come from? (c) Which country's athlete has the second best jump?
Australia	12.93	
Papua New Guinea	13.67	
New Zealand	12.69	
UK	13.08	
China	13.62	

Sports draws

Look carefully at these sporting draws and then answer the questions about them in your exercise book.

VOLLEYBALL
National Capital District
Saturday 11 March
Sir John Guise Stadium

8:00	Mixtures	vs	NCDC	WAR
9:30	Elvai	vs	Hoppers	WA
11:00	Scorpions	vs	Raukele	WA
12:30	NCDC	vs	Mixtures	WA
2:00	Telikom	vs	U-Miyet	WA
Court 2				
8.00	Telikom	vs	Hoppers	MAR
9:30	Mixtures	vs	Raukele	MA
11:00	NCDC	vs	Telikom	MA
12:30	Hoppers	vs	U-Miyet	MA
2:00	Elvai	vs	Scorpions	MA
Court 3				
8:00	U-Miyet	vs	Elvai	WAR
9:30	Raukele	vs	Telikom	WAR
11:00	Scorpions	vs	Hoppers	WAR
12:30	Elvai	vs	Scorpions	MAR
2:00	NCDC	vs	Raukele	MAR
3:30	U-Miyet	vs	Mixtures	MAR

RUGBY TOUCH
City
Sunday 12 March
Murray Barracks

12:00	Victims	vs	Defence	MIX
12:00	Magpies	vs	Dogs	MIX
12:30	Razorbacks	vs	Roosters	MIX
12:30	Diggers	vs	Hyundai	MIX
1:00	Konebypass	vs	Eagles	MIX
1:00	Red Ants	vs	Rebel Mix	MIX
1:30	Victims	vs	Defence	MIX
1:30	Magpies	vs	Dogs	W
2:30	Razorbacks	vs	Roosters	W
2:30	Diggers	vs	Hyundai	W
3:00	Konebypass	vs	Eagles	W
3:00	Red Ants	vs	Rebel Mix	W
3:30	Victims	vs	Defence	M
3:30	Magpies	vs	Dogs	M
4:00	Razorbacks	vs	Roosters	M
4:00	Diggers	vs	Hyundai	M
4:30	Konebypass	vs	Eagles	M
4:30	Red Ants	vs	Rebel Mix	M

1. Which teams are scheduled to play Rugby Touch at 3:30?
2. Which two teams are to play volleyball on Court 2 at 2 o'clock?
3. What time are the mixed Rugby Touch teams, Razorbacks and Roosters playing?
4. What is the earliest men's Rugby Touch game?
5. Which court has 6 volleyball matches scheduled?
6. How many volleyball matches did the Scorpions play on Saturday 11th March? What time were they?

Darts scores

Copy each table into your exercise book. Find the total of the throws and subtract it from 501 to find out what each player has left to finish.

1.

James		
Throws	Score	To finish
T20		
D16		
12		
D7		
T10		
18		
3		

2.

Emoto		
Throws	Score	To finish
15		
T18		
D20		
T4		
16		
D17		
11		

3.

Karl		
Throws	Score	To finish
D8		
9		
T13		
50		
D20		
5		
7		

4.

Bulima		
Throws	Score	To finish
D15		
T20		
T20		
50		
17		
D19		
T16		

5.

Gana		
Throws	Score	To finish
T14		
D9		
D16		
20		
50		
T19		
D6		

6.

Renagi		
Throws	Score	To finish
25		
T11		
D12		
T17		
6		
D14		
T20		

Work out some possible finishes for each player.

T=Triple
D=Double

Football scores

A In **Rugby** the scores work as follows:

try	**4 points**
conversion	**2 points**
field goal	**1 point**
penalty goal	**2 points**

1. What would be the total scores for a team that scored:

(a) 3 tries, 2 conversions, 2 field goals, and 1 penalty
(b) 4 tries, 1 conversion, 3 field goals, and 3 penalties
(c) 6 tries, 2 conversions, 2 field goals, and 2 penalties
(d) 8 tries, 6 conversions, 5 field goals, and 1 penalty
(e) 3 tries, 3 conversions, 1 field goal, and 2 penalties
(f) 6 tries, 4 conversions, 1 field goal, and 1 penalty?

2. At the end of a game a rugby team has scored 20 points. How many different ways might they have got this score?

B In **Australian Rules football**, the scores are as follows:

goal	**6 points**
behind	**1 point**

1. What would be the totals for teams that scored:

(a) 8 goals, 6 behinds
(b) 7 goals, 2 behinds
(c) 12 goals, 3 behinds
(d) 7 goals, 15 behinds
(e) 13 goals, 13 behinds
(f) 8 goals, 16 behinds
(g) 9 goals, 8 behinds
(h) 15 goals, 9 behinds
(i) 8 goals, 12 behinds
(j) 10 goals, 13 behinds
(k) 15 goals, 15 behinds
(l) 11 goals, 5 behinds?

2. If a team scored 100 points, how many different ways could they have done this?

3. Which team won this match?

Melbourne 9 goals, 13 behinds
Collingwood 10 goals, 5 behinds